The IUCN Species Survival Commission

2000 IUCN Red List of Threatened Species™

Compiled by
Craig Hilton-Taylor

SPECIES SURVIVAL COMMISSION

CONSERVATION INTERNATIONAL
CENTER FOR APPLIED BIODIVERSITY SCIENCE

CENTER FOR MARINE CONSERVATION

WWF

BirdLife INTERNATIONAL
Together for birds and people

Chicago Zoological Society

The IUCN Species Survival Commission

2000 IUCN Red List of Threatened Species™

Compiled by Craig Hilton-Taylor

with the assistance of Caroline Pollock, Matthew Linkie, Alain Mauric, Janice Long, Mariano Gimenez Dixon, Simon Stuart, Alison Stattersfield, Martin Sneary and Georgina Mace in association with experts in the IUCN Species Survival Commission and BirdLife International

Foreword by David Brackett
A Challenge to the Global Community by Russell Mittermeier

IUCN – The World Conservation Union
2000

This publication has been made possible in part by funding from SSC and Conservation International.

Published by: IUCN, Gland, Switzerland and Cambridge, UK in collaboration with Conservation International.

Citation: Hilton-Taylor, C. (Compiler) (2000). *2000 IUCN Red List of Threatened Species*. IUCN, Gland, Switzerland and Cambridge, UK. xviii + 61 pp.

ISBN: Book: 2-8317-0564-9
CD: 2-8317-0565-7

Cover design by: IUCN Publications Services Unit, Cambridge, UK

Layout by: Bookcraft Ltd, Stroud, UK

Produced by: IUCN Publications Services Unit, Cambridge, UK

Printed by: Thanet Press Ltd, Margate, UK

CD Design: Irene R. Lengui, L'IV Communications, Morges, Switzerland

CD produced by: Cymbiont Incorporated, Canada

CD printed by: Loadplan Ltd, London, UK

Available from: IUCN Publications Services Unit
219c Huntingdon Road, Cambridge CB3 0DL, United Kingdom
Tel: +44 1223 277894, Fax: +44 1223 277175
E-mail: info@books.iucn.org
http://www.iucn.org
A catalogue of IUCN publications is also available

The text of this book is printed on Fineblade Extra 90gsm made from low chlorine pulp

Contributions to the IUCN/Species Survival Commission and the *2000 IUCN Red List of Threatened Species*

The IUCN/Species Survival Commission gratefully recognizes its extensive network of volunteers who make production of the *IUCN Red List* possible. Those individuals who have contributed time and expertise are listed in the Acknowledgements. SSC also wishes to acknowledge those donors whose major financial contributions support a wide variety of SSC activities, as well as development and production of the *IUCN Red List*.

The Department of the Environment Transport and the Regions (DETR), UK is the IUCN State Member in the United Kingdom. The DETR supports the IUCN/SSC Red List Programme by financing the day-to-day running activities of the programme at the SSC centre in Cambridge, UK. This funding helps support the Red List Programme Officer and an intern to help run the programme. Together with two other UK Government-funded agencies, Scottish Natural Heritage and the Royal Botanic Gardens, Kew, the DETR is also financing a specialist plants officer.

Conservation International (CI) has generously provided the funds to help produce the *2000 IUCN Red List*, especially the production of a CD-ROM version of the list, the completion of the Red List component of IUCN's new internet-based World Conservation Atlas, and this publication. CI helps people improve their standard of living while conserving their valuable natural resources. CI develops and promotes models to conserve ecological "hotspots", threatened rain forests, and other ecosystems in Latin America, Africa and Asia. To ensure lasting solutions to conservation challenges, these models integrate economic, cultural and ecological factors, and are designed to strengthen local capacity for conservation.

The Center for Biodiversity Science (CABS), although part of CI, has a separate Executive Board. As a knowledge-based early warning system, CABS identifies critical issues confronting the conservation of biological diversity. It both anticipates destructive situations and is a pre-emptive force. The Center's mission is to strengthen our ability to respond rapidly, wisely and effectively to emerging threats to the Earth's biological diversity. To accomplish this, the Center mobilizes science, acts strategically, leverages partnerships, and broadens public outreach. CABS provided a fellowship for the Red List Programme Officer and generously supports in-kind two programme officers responsible for co-ordinating the activities of the SSC Primate and the Tortoise and Freshwater Turtle Specialist Groups. CABS, has also supported work to document the extinct birds which is incorporated in the 2000 Red List.

BirdLife International compiles and maintains the bird component of the *IUCN Red List*. Founded in 1922 under its original name of the International Council for Bird Preservation, BirdLife is a global conservation federation with a world wide network of partner organizations, representatives and dedicated individuals. BirdLife seeks to conserve all wild bird species and their habitats. Through this, BirdLife helps protect the world's biological diversity and supports the sustainable use by humans of the world's natural resources. BirdLife is a national force in 58 countries around the world and represented in a further 26 nations.

The Canadian Wildlife Service (CWS) of Environment Canada handles wildlife matters that are the responsibility of the Canadian government. These include protection and management of migratory birds as well as nationally significant wildlife habitat. Other responsibilities are endangered species, control of international trade in endangered species, and research on wildlife issues of national importance. CWS co-operates with the provinces, territories, Parks Canada, and other federal agencies in wildlife research and management. The CWS provides support to the SSC Red List Programme.

Natural Resources Canada (NRCan) is a federal government department specializing in energy, minerals and metals, forests and earth sciences. NRCan deals with natural resource issues that are important to Canadians, looking at these issues from both a national and international perspective, using its expertise in science and policy. How land and resources are managed today will determine the quality of life for Canadians both now and in the future. The GeoInnovations Fund of NRCan has funded the development of the internet-based *IUCN World Conservation Atlas*, which incorporates the 2000 *IUCN Red List* as one of its first products.

The Center for Marine Conservation (CMC), with its headquarters in the US, provides valuable in-kind funding and support to the marine work of SSC. It is the major funder of the SSC Marine Turtle Specialist Group (MTSG), employs the MTSG Programme Officer, and administers funds on behalf of the SSC Shark and Cetacean Specialist Groups. A CMC staff member acts as SSC staff liaison for the marine specialist groups and as the marine focal point for SSC, and also supports the development of SSC's work in the marine realm. CMC serves as the marine focal point for the IUCN/SSC Red List Programme. It is dedicated to protecting ocean environments and conserving the global abundance and diversity

The Nature Conservancy (TNC) is an international conservation organization dedicated to preserving the plants, animals, and natural communities that represent the diversity of life on Earth by protecting the lands and waters they need to survive. With more than one million members, the Conservancy has protected over ten million acres in the United States alone, and owns and manages the largest network of private nature reserves in the world. TNC supplied data for many of the North American species included in the Red List.

The Association for Biodiversity Information (ABI) is a non-profit organization dedicated to managing and distributing authoritative information critical to the conservation of the world's biodiversity. Working in partnership with the TNC network of natural heritage programmes and conservation data centres, ABI strives to create regional- and national-level products that promote the use of sound scientific information in environmental decision-making. ABI was formerly the data management and scientific wing of TNC, and as such, will in future be the main provider of information on many of the North American species on the Red List.

The World Wide Fund for Nature (WWF) provides significant annual operating support to the SSC. WWF's contribution supports the SSC's minimal infrastructure and helps ensure that the voluntary network and Publications Programme are adequately supported. WWF aims to conserve nature and ecological processes by: (1) preserving genetic, species, and ecosystem diversity; (2) ensuring that the use of renewable natural resources is sustainable both now and in the longer term; and (3) promoting actions to reduce pollution and the wasteful exploitation and consumption of resources and energy. WWF is one of the world's largest independent conservation organizations with a network of National Organizations and Associates around the world and over 5.2 million regular supporters. WWF continues to be known as World Wildlife Fund in Canada and in the United States of America. WWF–UK provided support to the IUCN/SSC Red List Programme.

The Council of Agriculture (COA), Taiwan has awarded major grants to the SSC's Wildlife Trade Programme and Conservation Communications Programme. This support has enabled SSC to continue its valuable technical advisory service to the Parties to CITES as well as to the larger global conservation community. Among other responsibilities, the COA is in charge of matters concerning the designation and management of nature reserves, conservation of wildlife and their habitats, conservation of natural landscapes, coordination of law enforcement efforts as well as promotion of conservation education, research and international cooperation.

The George B. Rabb IUCN/SSC Internship Programme: Dr George B. Rabb is widely recognized as one of the most influential thinkers on conservation and environmental issues during the latter part of the 20th Century. Dr Rabb is President of the Chicago Zoological Society, and, for seven years, served as Chair of the IUCN Species Survival Commission. When Dr Rabb stepped down as Chair of the SSC in 1996, the members of the Commission decided to commemorate his outstanding leadership by establishing an endowment to pay for interns from around the world to work on key projects for the SSC. The George Rabb internship is a prestigious, competitive award for young professionals seeking a long-term career in conservation. The first George Rabb Intern appointment was for a six-month period, running from April through September 2000, to help with the compilation of the *2000 IUCN Red List*.

The Chicago Zoological Society (CZS) provides significant in-kind and cash support to the SSC, including grants for special projects, editorial and design services, staff secondments and related support services. The mission of the CZS is to help people develop a sustainable and harmonious relationship with nature. The Zoo carries out its mission by informing and inspiring two million annual visitors, serving as a refuge for species threatened with extinction, developing scientific approaches to manage species successfully in zoos and the wild, and working with other zoos, agencies, and protected areas around the world to conserve habitats and wildlife.

The UNEP World Conservation Monitoring Centre (UNEP-WCMC) was established in 2000 as the world biodiversity information and assessment centre of the United Nations Environment Programme. The roots of the organization go back to 1979, when it was founded as the IUCN Conservation Monitoring Centre. In 1988 the World Conservation Monitoring Centre was created jointly by IUCN, WWF-International and UNEP. The Centre's activities include assessment and early warning studies in forest, dryland, freshwater and marine ecosystems. Research on endangered species and biodiversity indicators provide policy-makers with vital knowledge on global trends in conservation and sustainable use of wildlife and their habitats. Extensive use is made of geographic information systems and other analytical technologies that help to visualize trends, patterns and emerging priorities for conservation action. UNEP-WCMC is committed to the principle of data exchange, and acts as a clearing-house allowing data providers and users to share data and information. Wherever possible, data managed by the Centre is placed in the public domain. WCMC was for many years SSC's data management partner and helped compile the *IUCN Red Lists*.

Contents

Foreword

On numerous occasions I have found reason to state my amazement at what the dedication of the IUCN Species Survival Commission (SSC) network of scientists and conservation specialists can accomplish. This dedication is abundantly demonstrated by the product you have before you, the *2000 IUCN Red List of Threatened Species*. The current analysis, which combines the assessment of both flora and fauna in one list, is another step forward in a progression which began with the Red Data Sheets many years ago. From those beginnings, the science of conservation has grown, as have the IUCN Red Lists.

The current Red List Programme began when the SSC launched an initiative to revise the listing system, in recognition of the need for a consistent and objective process to describe threatened species. Through work lead by Georgina Mace, a quantitative system of categories and criteria for listing were adopted by the IUCN Council in 1994. The adoption of IUCN Red List Categories ushered in a new age of rigorous assessment and of national awareness. A body of scientific literature based on the review and analysis of the IUCN Categories and Red Lists was born. In addition, a burgeoning awareness of an extinction crisis was recognized at a national level as many countries adopted the IUCN Red List system to form the basis for national red lists.

Within the SSC, the development of a quantitative system led to the creation of the Red List Programme, and to the book before us. There now exist Red List Authorities, responsible for ensuring that species are evaluated against the IUCN Red List Categories in a peer-reviewed manner. Further, the Red List Programme is moving towards clarifying the rationale for species listing with improved documentation and taxonomic standards. The consequent data underlying the fully documented Red List will be maintained within the SSC Species Information System. The advent of appointed Red List Authorities, improved documentation and validated taxonomy will continue to make the listing process more transparent and thus more open to challenge. Intellectual scrutiny and challenge is at the heart of peer-reviewed science and can only increase the scientific integrity and authority of the IUCN Red Lists. A petitions process has been established to formally appraise listing challenges.

The ongoing evolution of the Red List Programme has set the stage for a meaningful long-term analysis and monitoring of biodiversity through the establishment of indicators of biodiversity trends. The current analysis is a momentous initial step in the assessment of global biodiversity.

The *1996 IUCN Red List of Threatened Animals* analysed all bird and mammal species against the IUCN Red List Categories. These species have been re-evaluated here, against the same categories, and the results are telling. In the *1996 IUCN Red List of Threatened Animals*, there were 169 Critically Endangered (CR) and 315 Endangered (EN) mammals; the current analysis now designates 180 CR and 340 EN mammals. Similarly for birds, there is an increase from 168 to 182 CR designated and 235 increased to 321 EN designated species in 2000. While the overall percentage of mammals and birds at risk (CR, EN or Vulnerable) has not greatly changed, the magnitude of risk, as represented by IUCN Red List Category, has increased. Clearly there is a documented deterioration of the status of species;

clearly there is an extinction crisis before us. What the impact of this continued crisis will be has yet to be fully realized.

However ominous the analysis, there is also a clear indication of where action in needed. The quantifiable extinction crisis, the future establishment of biodiversity trend indicators, and the continued adaptation of the Red List Programme all look towards determining the underlying causes of the crisis in the hope that viable solutions will be developed. Fast action is needed, but the tools are in hand.

The formidable task now prioritized by the Red List Programme will be to fully expand this analytical skill to the less well-represented taxa. It is recognized that mammals and birds alone may not provide the best indication of the global biodiversity situation. There is a need to incorporate other taxa for a more comprehensive analysis. This is a significant challenge, one that I believe the intellectual force of the network is equipped to confront with the support of the Red List Programme.

When I consider the current achievements and the future challenges, all of which began with a handful of Red Data sheets and a network of committed scientists and conservation specialists, is it any wonder I continue to be amazed?

David Brackett
Chair, IUCN Species Survival Commission

A challenge to the global community

I still have fond memories of receiving in the mail my copy of the first *Red Data Book*, the Mammal volume compiled by Noel Simon and published in 1966. It was in a small format, red loose-leaf binder, with the endangered species being given pink sheets, the vulnerable species yellow sheets and the rare species white sheets. Later, it was periodically updated with a stack of new loose-leaf sheets that could be inserted into the binder. I was about 20 when I first received this publication, and it had a profound impact on me. I pored over every page, reading each one dozens of times, feeling awful about those species that were severely endangered, and resolving to dedicate my career to doing something on their behalf. A budding primatologist, I especially remember the primate sheets, among them the Golden Lion Tamarin, the Indri, the Aye-aye, and the Lion-tailed Macaque, and was impressed by how little we knew about these wonderful creatures and how they cried out for more attention. In the early years of the 1970s, I travelled to South America with the express intent of assessing the status of some of these species, among them the red, white and black uakaris, the White-nosed Saki, the Yellow-tailed Woolly Monkey, and later the Lion Tamarins and Muriqui – the first steps in what became a 30-year career working on the conservation of Neotropical monkeys.

To say that the first *Red Data Book* helped to define my career would be an under-statement, and I know that many of my colleagues were similarly affected. What is more, this first publication on Mammals, and the two others on Birds (1968) and Reptiles and Amphibians (1968) that followed shortly thereafter, stimulated the production of dozens of subsequent global, regional and national *Red Data Books* in the 1970s and 1980s and the enactment of a host of endangered species laws around the world. This process had enormous catalytic impact, and must be rated as one of the most significant achievements in the history of biodiversity conservation.

At the same time, I remember how fragile and how anecdotal most of the data on endangered species actually was in the early days. In particular, I remember back in 1974, sitting down with Brazilian conservation pioneer, Dr. Adelmar Coimbra-Filho, in the Biological Bank for Lion Tamarins in Rio de Janeiro and revising the Brazilian mammals for an update of the *Mammal Red Data Book.* We added new species to the list, changed the status of several others, and substantially increased the amount of information available for Brazil. However, virtually across the board, the information that we provided was anecdotal, sometimes based on a small handful of field observations, but not including a single long term-study of any of the listed species. Whether the species was endangered or vulnerable was really nothing more than "expert" opinion, a guess often based on the most fragmentary information. Nonetheless, as imperfect as it all was, it represented "state-of-the-art" at that time, and had enormous impact. Brazil subsequently passed a host of endangered species laws, increased the area under protection a hundred fold, and has produced a series of its own *Red Data Books* at national and even state level. The same has been true in many other countries.

Needless to say, we have come a long way since those early days. Beginning in the late 1980s, it was recognized that we needed much more rigorous criteria for assessing conservation status, and that these criteria had to be defensible in the strongest scientific terms. Furthermore, with the explosion of fieldwork on endangered species in the last three decades of the 20th century, much more information became available, especially on the previously very poorly known biodiversity of the tropics. With the pioneering work by Georgina Mace and Russ Lande, published in 1991, and the follow-up studies by Mace and colleagues, we now have an excellent set of quantitative criteria for assessing status. We have also increased the number of categories to reflect the many subtleties of this process, including the very important category of *Critically Endangered* for those species that are truly on the edge. These criteria were first applied at a global level in the 1996 Red List, which was launched at the World Conservation Congress in Montreal in 1996 and marked the beginning of a new era in efforts to conserve endangered species. Over the past few years, we have seen continued refinement of the criteria, and the establishment of a formal Red List Programme and a Red List Subcommittee involving some of the world's leading experts on this issue. We have also seen an increased willingness to invest in this endeavor, a strong indication that the international conservation community is finally recognizing that this process is perhaps *the* single most fundamental component of the biodiversity conservation movement.

Our organization, Conservation International, has long recognized the critical importance of the Red-listing process, and has helped to support this over the past decade. This has included underwriting various national and taxon-based Red Data Books and Red Lists, co-sponsorship with SSC of the *1996 IUCN Red List*, and major support for the Red List Programme through our Center for Applied Biodiversity Science (CABS), beginning in 1999. Now, we are pleased once again to co-sponsor the production of the 2000 Red List with SSC. Our commitment to this work grows with each passing year, and we will certainly support it even more vigorously in the future.

Finally, a word about the urgency of the global extinction crisis. This has been much discussed in recent years and is starting to be given the international attention that it so rightly deserves. However, once again, this new Red List highlights how many wonderful creatures could be lost in the first few decades of the 21st century if we, as a global society, do not greatly increase our levels of support, involvement and commitment. Although the total number of listed species has not increased dramatically from 1996 to 2000 (e.g. mammals go from 1096 to 1130 and birds from 1107 to 1183), the fact that the number of Critically Endangered species has increased (mammals from 169 to 180; birds from 168 to 182) is cause for much concern. Even more frightening is the increase in the number of bird extinctions, from 108 recognized in 1996 to 131 in the 2000 list. Although this to some extent reflects improved documentation, the loss of any species should be taken very seriously indeed by the global community.

Another very important point is that Red List species are by no means evenly distributed over the surface of the planet. They tend to be concentrated in certain countries and certain ecoregions that are particularly rich in endemic species and have been most heavily impacted by our own species. This is best demonstrated by the 25 biodiversity "hotspots", which have been Conservation International's major theatre of operation for the past 11 years. These areas have enormous concentrations of endemic species and are under severe threat, having together already lost 88% of their original extent. In the small area that remains, amounting to only 1.4% of the land surface of the planet, we have concentrated as endemics fully 44% of all plants and 35% of all non-fish vertebrates. Not surprisingly, more than two-thirds of the world's most endangered mammals and more than 80% of the most endangered birds come from these hotspots.

Some of these hotspots really stand out in terms of their fragility and their concentrations of Red List species. The Atlantic Forest Region of Brazil has already lost 93% of its original natural vegetation and has very high numbers of endangered primates, birds, and orchids. The Mesoamerica Hotspot, encompassing southern Mexico and most of Central America, is another example. And the Indo-Burma Hotspot, including Indo-China, Thailand, Burma and extreme northeastern India, has both the highest diversity of turtles on Earth and the largest number of endangered turtles – in large part due to a runaway, uncontrolled food and medicinal trade to China. The list goes on.

The same is true in terms of priority countries. The megadiversity countries that we defined in a 1997 publication total 17 in number and have within them at least two-thirds of global biodiversity – terrestrial, freshwater and marine. Not surprisingly, they rank high on the list of countries with the most endangered mammals, birds and plants. For example, of the top 20 countries for listed mammals, 14 are megadiversity countries, with the world's two biologically richest countries, Indonesia and Brazil, coming in first and third in terms of largest number of threatened mammals.

Madagascar is one of the classic examples. Both a hotspot and a megadiversity country, this fantastic island has extremely high levels of endemism but has already lost more than 90% of its original natural vegetation. It has more Critically Endangered and Endangered primates than anywhere else, not to mention an entire primate megafauna of at least 8 genera and 15 species that has already gone extinct in the past two millennia. The Philippines, another island nation that is both a top priority hotspot and a megadiversity country, has lost 97% of its original vegetation and has more Critically Endangered birds than anywhere else.

What is the message from this new 2000 Red List? In my opinion, this new list demonstrates quite clearly that we are in an extinction crisis. The fact that the number of Critical and Endangered species has increased and that many of the 1996 Endangered species have moved into the Critical category indicates that we are sitting on a time bomb – even if the lag effects make it likely that the full impacts will not be felt immediately. The message in all of this is quite simple. We need to act decisively, we need to act now, and we need to act at a scale far beyond anything that has ever been done before. The findings in the 2000 Red List, together with conservation priority setting exercises like the hotspots and megadiversity countries mentioned above, provide us with a solid foundation as to where we need to focus. We now need to mobilize the human and financial resources at a level at least one and more likely two orders of magnitude beyond anything previously realized. In the past, this may have seemed wonderfully idealistic and totally unrealistic, but it is now within our grasp. We have to join forces with a wide range of partners, continue to develop strong relationships with governments and local communities, and, most important of all, engage an increasingly interested private sector at a new level.

At the same time, I think that we have to use the 2000 Red List and all the other results that emerge from our work to generate much more public support for and interest in the extinction crisis. Global society would be horrified if someone set fire to the Louvres in Paris or the Metropolitan Museum in New York, or if someone blew up the Pyramids or the Taj Majal. Yet every time a forest is burned to the ground in Madagascar or the Philippines, the loss to global society is at least as great, yet no one pays very much attention – and sadly it happens every day. Every time a species goes extinct, every time a study reveals that a species has entered into the Critically Endangered category, every time a unique habitat is destroyed, we have to make sure that the world knows and we have to make the same fuss that we would if the loss were one of our own human creations. If we can do this, if we can make the loss of every species a cause for global mourning, if we can change our personal and societal values to better reflect the importance of the full range of life on Earth, I think we could be well on our way to achieving success. The Red Books and Red Lists have already made a major contribution to our understanding of the extinction crisis over the past 35 years. We now need to use this and future lists and the Red List Programme as a whole to take this issue to the next level and make it one of the highest priorities for global society in the decade to come.

Russell A. Mittermeier
President, Conservation International
and Chairman, IUCN/SSC Primate Specialist Group

Acknowledgements

The compilation and production of an IUCN Red List such as this would not be possible without the extraordinary enthusiasm, dedication and willingness of many people world-wide to contribute information and much of their valuable time. In particular, we must acknowledge the many SSC Specialist Group chairs, Red List Authority focal points, Specialist Group members, and the many field scientists, who have provided data and opinions for this revision. In the list below we have tried to highlight those Specialist Groups (in alphabetical order) and individuals within each group who have actively contributed this time, but we also need to draw attention to the fact that the acknowledgements to all the people who contributed to the *1996 IUCN Red List of Animals* and *The World List of Threatened Trees* are equally valid here, as the 2000 Red List is in part a compilation of everything that appeared in these two publications. In the list below, it is often only the chair or Red List focal point from a Specialist Group who is mentioned. They are in fact representative of a larger number of members in each Specialist Group who actively contribute to the Red List and we wish to recognize the valuable contributions of all these unnamed people.

African Rhino Specialist Group (SG) – M. Brooks, R.H. Emslie; **Antelopes SG** – R. East, P. Grubb, S. Kingswood, D. Mallon; **Asian Wild Cattle SG** – J.W. Duckworth, S. Hedges; **Australasian Marsupials & Monotremes SG** – A.A. Burbidge; **Australasian Reptile & Amphibian SG** – C.B. Banks, H.G. Cogger; **BirdLife International** – D. R. Capper, N. J. Collar, M. J. Crosby, G. C. L. Dutson, M. I. Evans, R. K. McClellan, N. B. Peet, S. M. Shutes, A. J. Stattersfield, T. E. H. Stuart, J. A. Tobias, D. C. Wege (all BirdLife Secretariat); K. Barnes (BirdLife South Africa); L. Bennun (Nature Kenya), D. Callaghan (BirdLife European Office); S. Chan (International Centre, Wild Bird Society of Japan); S. Garnett and G. Crowley (Birds Australia); A. F. A. Hawkins (BirdLife International Madagascar Country Programme); J. Carroll (Partridge, Quail and Francolin SG); J. P. Croxall (British Antarctic Survey); R. Dekker (Megapode SG); E. H. Dunn (Canadian Wildlife Service); P. Garson (Pheasant SG); P. McGowan (World Pheasant Association); G. Sherley (South Pacific Regional Environment Programme); G. Taylor and H. Robertson (Department of Conservation, New Zealand); **Bison SG** – Z. Pucek; **Bryophyte SG** – P. Geissler, T. Hallingbäck, N. Hodgetts, G. Raemaekers, R. Schumacker, C. Sérgio, L. Söderström, N. Stewart, P. Tan, J. Vána; **CBSG CAMP India Workshop participants** – N.V.K. Ashraf, R. Borges, S. Chakraborty, A.K. Chakravarthy, S. Chattopadhyay, G. Christopher, J.C. Daniel, P.S. Easa, E.A. Jayson, G.K. Joseph, R. Krishnan, R.S. Lal Mohan, M.M. Mansoor, G. Marimuthu, V. Menon, M. Mishra, S. Molur, D. Mudappa, K. Mukherjee, M. Muni, P.O. Nameer, M.S. Pradhan, S. Paulraj, K.K. Ramachandran, G. Ramaswamy, M.V. Ravikumar, Y.P. Sinha, S. Sivaganesan, K. Srihari, K.A. Subramanian, W. Sunderraj, S. Varman, A. Venkataraman, S. Walker (Indian Endemic Mammals); **Cacti & Succulent SG** – W. Stuppy; **Caprinae SG** – I.L. Alados, R.B. Harris, S. Lovari, D. Shackleton, Yu Yuqun; **Carnivorous Plant SG** – B. von Arx, R. Cantley, P.M. Catling, M. Cheek, C. Clarke, J.G. Conran, G. Folkerts, C. Frost, R. Gardner, M. Groves, M. Jebb, G. Leach, A. Lowrie, L. Mellichamp, J. Nerz, P. Perret, H. Rischer, J.H. Rondeau, J. Schlauer, D.E. Schnell, A. Wistuba; **Cat SG** – P. Jackson; **Cetacean SG** – R.L. Brownell, Jr., D. Kreb, R. Reeves, L. Slooten, B.D. Smith, B. Taylor, D. Weller; **Chiroptera SG** – W. Bergmans, A.M. Hutson, S.P. Mickleburgh, P.A. Racey; **Conifer SG** –

J.A.Bartel, A. Farjon, C.N. Page, T. Yahara; **Crocodile SG** – J. Perran Ross; **Declining Amphibian Population Task Force** – A.C. Alcala, A.C. Diesmos, S.L. Kuzmin; **Freshwater Fish SG** – L. Kaufman, P. Skelton; **Grouper & Wrasse SG** – G. Garcia-Moliner, D. Pollard, Y. Sadovy, M. Samoilys; **Hyaena SG** – H. Hofer, G. Mills; **Iguana SG** – A. Alberts, J. Bendon, D. Blair, R. Carter, M. Day, M. Garcia, G. Gerber, W. Hayes, J. Iverson, N. Perez, J. Wasilewski, T. Wiewandt; **Insectivore, Tree Shrew & Elephant Shrew SG** – R. Hutterer; **Indian Ocean Island Plant** SG – M. Allet, V. Bachraz, M.E. Duhloo, D. Florens, R. Rutty, W. Strahm, V. Tezoo; **Marine Turtle SG** – A. Abreu, D. Crouse, A.L. Sarti Martínez; **Mollusc SG** – M. Baker, A.E. Bogan, K. Bonham, Helsdingen, M.C. Mansur, M.B. Seddon; **Mustelid, Viverrid, & Procyonid SG** – L. Dollar, H. van Rompaey, R. Wirth; **Odonata SG** – N. Moore; **Otter SG** – J.A. Estes, A. Fischotterschutz, R. Green, A. Gutleb, S.A. Hussain, B. Kanchanasakha, P. Kumari de Silva, G. Medina Vogel, J.A.J. Nel, C. Reuther, C. Schenck, T. Serfass, N. Sivasothi; **Pigs, Peccaries & Hippoes SG** – W.L.R. Oliver; **Primate SG** – S.K. Bearder, C. Beth-Stewart, D. Brandon-Jones, T.M. Butynski, L. Cortes-Oritz, W. Dittus, A. Eudey, J. Ganzhorn, T. Geissmann, K.G. Glander, D.A. Hill, W.R. Konstant, A. Langguth, D. Melnick, S.L. Mendes, R.A. Mittermeier, J. Oates, E. Rodriguez-Luna, A.B. Rylands, M. Shekelle, H. Schneider, T.T. Struhsaker, I. Tattersall; **Rodent SG** – G. Amori, D. Belitsky, V.C. Bleich, B.R. Blood, C.J. Conroy, J.A. Cook, B. Csuti, J.R. Demboski, J.R. Easton, R.A. Erickson, R.A. Fridell, D.J. Hafner, J.C. Hafner, J.G. Hall, J.H. Harris, T. Kawamicki, K.A. Kime, G.L. Kirkland, Jr., E. Weintraub Lance, W.Z. Lidicker, Jr., S.O. MacDonald, R.E Martin, K.G. Matocha, K.P. McDonald, C. McGaugh, S. Mullin, S.J. Myers, D.W. Nagorsen, M.J. O'Farrell, M.A. Patten, B. Peyton, I. Prakash, A.I. Roest, H.S. Shellhammer, R.B. Spicer, D.T. Steele, J.M. Sulentich, R.M. Sullivan, C.S. Thaeler, Jr., E. Yensen; **Shark SG** – W.F. Adams, A.F. Amorim, M.L.G. Araújo, C.A. Arfelli, R. Bonfil, S. Bransetter, G.H. Burgess, G.M. Cailliet, M. Camhi, J. Casey, G.E. Chiaramonte, G. Cliff, L.J.V. Compagno, C. Conraths, S.F. Cook, E. Cortés, J. Denham, M. Drioli, N. Dulvey, J. Ellis, L. Fagundes, I. Fergusson, S. Fordham, S.L. Fowler, M.P. Francis, D. Gledhill, K.J. Goldman, S. Gruber, M. Heupel, B. Human, H. Ishihara, J.E. Kotas, M. Krose, P. Last, R. Lessa, M. Marks, R. Myers, J.A. Musick, J.A. Neer, B. Norman, G. Notarbartolo di Sciara, M.I. Oetinger, M. Pawson, D. Pollard, R.S. Rosa, C. Simpfendorfer, M.J. Smale, A. Smith, S.E. Smith, J. Stevens, L.F. Sundström, C.M. Vooren, P. Walker, T. Walker, G. Zorzi; **Tapir SG** – C.C. Downer, S. Todd; **Southern African Plant SG** – J. Golding, M.F. Mfab, J. Victor; **Tortoise & Freshwater Turtle SG and the Asian Turtle Trade Working Group** – J.L. Behler, S. Bhupathy, B. Chan, T-h. Chen, B.C. Choudhury, I. Das, P.P. van Dijk, D. Hendrie, D. Iskandar, R. Kan, S.M. Munjurul Hannan Khan, M. Lau, H. Ota, T. Palasuwan, S. Platt, S.M.A. Rashid, A. Rhodin, D. Sharma, C. Shepherd, H-t. Shi , B. Stuart, R. Timmins and Y. Yasukawa.

The following people who are not members of any of the above mentioned groups also provided new assessments (in some cases substantial contributions) and/or new information:

P. Achery, L.C. Afuang, J. Aldaz, A.J.S. Argolo, T. Bell, R.S. Bérnils, B. Bolton, M. Brendell, D. Brummitt, S. Cable, M.A. Carretero, D.S. Carroll, K.M. Catley, P. Choate, M. Cheek, M. Cherry, J. Davies, A.M.H. de Bruyn, M. Di Bernardo, R.C. Dowler, C.W. Edwards F.L. Franco, J.C. Gonzales, D. Goodger, J.C. Guix, A. Guttman, N. Grange, D.S. Hammond, T.Hecht, N. Ishii, J. Jarvie, P. Joubert, B. Lang, G.A. Llorente, J. Lockyear, C. Lyal, O.A.V. Marques, J. Marshall, M.B. Martins, A. Mauchamp, S.D. Mikhov, A. Montori, S.A.A. Morato, J.C. Moura-Leite, R.B. Oliveira, E. Ortiz, J. Ottenwalder, B. Pollard, J. Reynolds, C.F.D. Rocha, A. Russell, X. Santors, K. Sattler, S. Shute, D. Telnov, A. Tye, H. Valdebenito, C. van Swaay, H. ter Steege, A. Vincent, B.J.W. van der Wollenberg and M. Warren.

The data on birds was provided by BirdLife International from their new publication *Threatened Birds of the World*, which is the official source for birds on the 2000 *IUCN Red List*. We would like to thank Alison Stattersfield, David Capper, Martin Sneary, Colin Bibby and Adrian Long for all their help in providing the bird information and for the valuable exchange of ideas and information during the compilation of the Red List.

Mary Seddon and Martin Baker are thanked for their efforts to improve the accuracy of the mollusc listings in the 2000 Red List. We are especially grateful to Mary Seddon for providing additional information on molluscs for use in the analysis and for providing valuable feedback on the design of the *IUCN Red List* web site.

The compilation of the 2000 Red List began in 1999 when it was planned to produce an interim 1999 update to the Red List on the World Wide Web only, which would combine information from the *1996 IUCN Red List of Threatened Animals* with that from *The World List of Threatened Trees* and whatever updates the SSC network could provide. The SSC negotiated a contract with the World Conservation Monitoring Centre (now UNEP – WCMC) to do this. For various reasons it was decided not to produce the 1999 List, but the information from the electronic data set compiled has formed the basis for this edition. WCMC are gratefully acknowledged for all their help in combining the data sets and updating the information. Particular thanks are due to Nick Teall and Tim Johnson for overseeing the process and to Stephen Grady who entered the new information and corrections into the database. Megan Cartin helped Craig Hilton-Taylor co-ordinate SSC's contribution to this process. The past experience and knowledge of UNEP – WCMC staff involved in the compilation of the previous editions of the Red List was invaluable, and particular thanks are due to Mary Cordiner, Neil Cox, Harriet Gillett, Brian Groombridge, Tim Inskipp and Charlotte Lusty.

Staff from Conservation International's Center for Applied Biodiversity Science (CABS) helped play a major co-ordinating role in gathering data to be used in this Red List. In particular we thank Thomas Brooks for providing all the extinction data on birds, Kurt Buhlmann for helping and advising on herpetofauna assessments and especially in getting the revised freshwater turtle assessments from Southeast Asia submitted, to Claude Gascon for help with amphibian assessments, and finally we are indebted to Anthony Rylands for liaising with the members of the Primate Specialist Group. We thank Gustavo Fonseca, Executive Director of CABS, for allowing his staff to dedicate so much of their time to the Red List.

The task of compiling the 2000 Red List of Threatened species presented a number of new challenges. For the first time in many years, SSC through the Red List Programme was undertaking the compilation itself, rather than contracting it out to another organization. To do this required the development of an interim database so that the information could be collated and analysed. Fabio Corsi from the SSC Species Information Service provided guidance and emergency help when required during this process. Wendy Strahm provided valuable input on the plant assessments submitted for Mauritius. The team at the SSC Cambridge office comprised Craig Hilton-Taylor (IUCN/SSC Red List Programme Officer), Matthew Linkie (The George B. Rabb Intern), Alain Mauric (IUCN/SSC Plants Officer) and Caroline Pollock (the Red List Programme Intern). This whole team together with additional support from Mariano Gimenez Dixon and Mandy Haywood worked beyond the call of duty to meet the very tight deadlines for this production. In spite of many tense moments, sleepless nights and database failures, good humour and camaraderie was maintained. Janice Long joined the team during the final phase to help with the analysis and the production of all the graphics used in the analysis. Georgina Mace was a constant source of wisdom and advice throughout the process. Susan Mainka read through the final text and provided editorial input. Simon Stuart was the IUCN project manager and served as overall editor of this publication.

Design and production was coordinated at the offices of the IUCN Publications Services Unit in Cambridge by Tiina Rajamets, Anne Rodford and Elaine Shaughnessy, and typesetting and layout undertaken by Bookcraft Ltd. The production of the web site and CD-ROM was due to the inspiration of Jean Thie, the IUCN Head of Information Management and the Canadian company Cymbiont Incorporated who were contracted to do the work. The Cymbiont development team comprised Charles MacLean, Eric Dewhirst, James Lee, Steve O'Brien, Adrian Irving-Beers, Sabina Kerchowski, Danny Brown and Adam Elliott. The CD-ROM was designed by Irene R. Lengui, L'IV Communications.

Production of the 2000 IUCN Red List of Threatened Species would not have been possible without the ongoing support of the long-term donors to the IUCN Species Survival Commission and to the Red List Programme in particular: Conservation International's Center for Applied Biodiversity Science; the Department of Environment, Transport and the Regions, UK; WWF – The World Wide Fund for Nature; the Canadian Wildlife Service; the Council of Agriculture, Taiwan; the Department of State, USA; the Center for Marine Conservation; Natural Resources Canada; the National Science Foundation (US); the Swiss Development Corporation; and the Chicago Zoological Society. In particular, the final production of this Red List on the web and on CD-ROM, and the publication of this book would not have been possible without generous funding from IUCN/SSC's co-publishing partner, Conservation International. Particular thanks are due to Russell Mittermeier for making all this possible.

Introduction

The magnitude and distribution of species that exist today is a product of more than 3.5 billion years of evolution, involving speciation, migration, extinction and, more recently, human influences. Estimates of the total number of species in existence range from 7 to 20 million, but perhaps the current best working estimate is between 13 and 14 million species (Hammond 1995). Collation of information on species from systematists and the taxonomic literature suggests that only 1.75 million of the species that exist have been described (Hammond 1995). There is now a growing world-wide concern about the status of this biodiversity, especially the loss of many undescribed species. It has been estimated that current species extinction rates are between 1,000 and 10,000 times higher than the background rate (May *et al.* 1995). Many species are declining to unsafe population levels, important habitats are destroyed, fragmented and degraded, and ecosystems are destabilized through climate change, pollution, invasive species, and direct human impacts. But there is also growing awareness of how biodiversity supports livelihoods, alleviates poverty, enables sustainable development, and fosters co-operation between nations.

Yet remarkably the world lacks a functioning system for monitoring the status and trends of biodiversity. Compared with almost every other sector (e.g., trade, financial flows, health, climate, etc.), biodiversity is only monitored in the most rudimentary manner. As a result, we know little about relative rates of biodiversity loss around the world, and even less about the vulnerabilities of species groups and ecosystems. We cannot assess the medium- and long-term effects of human activities on biodiversity. The *IUCN Red List* is being developed as one of the tools that will help to assess and monitor the status of biodiversity.

IUCN–The World Conservation Union, through its Species Survival Commission (SSC) has been producing Red Data Books and Red Lists for almost four decades. However, the publication of the *1996 IUCN Red List of Threatened Animals* (Baillie and Groombridge 1996) marked a major turning point in this long history. For the first time in a single global list, the conservation status of all species of birds and mammals (rather than just the better-known or more charismatic species) was evaluated, and all the assessments were based on the then new quantitative criteria introduced by IUCN in 1994 (IUCN 1994, see Annex 6). The result was a much more comprehensive and systematic treatment of two whole classes of organisms than had previously been available, which enabled reliable comparisons to be made. Conclusions about the status of the world's biodiversity could now be much more easily substantiated with sound scientific evidence.

The 1996 turning point was prompted in part by the development of the 1994 *IUCN Red List Categories* and Criteria, but also to a large degree by the high standards being set by IUCN's partner organization – BirdLife International. Birds are by far the best-known group of organisms on the planet, with a relative wealth of distribution and population data available enabling BirdLife to produce a global analysis in *Birds to Watch 2* (Collar *et al.* 1994), thereby establishing a new standard yet to be matched for most other groups of organisms.

Since their inception, the IUCN Red Data Books and Red Lists have enjoyed an increasingly prominent role in guiding conservation activities of governments, NGOs and scientific institutions. The *IUCN Red Lists* are widely recognized as the most comprehensive, apolitical global approach for evaluating the conservation status of plant and animal species. From their small beginnings, the *IUCN Red Lists* have grown in size and complexity. The introduction in 1994 of a scientifically rigorous approach to determine risks of extinction which is applicable to all species and infra-specific taxa, has become a virtual world standard (WCMC 2000). In order to produce Red Lists of all threatened species world-wide, the SSC has to draw on and mobilize a network of scientists and partner organizations working in almost every country in the world, who collectively hold what is likely the most complete scientific knowledge base on the biology and conservation status of species. The process for achieving this was largely uncoordinated and opportunistic. As a result, in 1998 the SSC Executive Committee agreed to the development of a coherent well-conceived Red List Programme with a management and governance plan that would ensure the highest standards of documentation, information management, training, and scientific oversight. The *IUCN Red List* Programme and its companion information management system (the Species Information Service) will provide fundamental baseline information on the status of biodiversity as it changes over time.

The **goals of the *IUCN Red List* Programme** are to:

- Provide a global index of the state of degeneration of biodiversity; and
- Identify and document those species most in need of conservation attention if global extinction rates are to be reduced.

The first of these goals refers to the traditional role of the *IUCN Red List*, which is to identify particular species at risk of extinction. The role of the *IUCN Red List* in underpinning priority setting processes for single species remains of critical importance. However, the second goal represents a radical new departure for the SSC and for the Red List Programme, for it focuses on using the data in the Red List for multi-species analyses in order to understand what is happening to biodiversity more generally.

To achieve these Goals, the following **Objectives** are proposed:

- To assess, in the long term, the status of a selected set of species;
- To establish a baseline from which to monitor the status of species;
- To provide a global context for the establishment of conservation priorities at the local level; and
- To monitor, on a continuing basis, the status of a representative selection of species (as biodiversity indicators) that cover all the major ecosystems of the world.

In relation to these goals and objectives it was agreed that the *IUCN Red List* should be characterized by the following **Operating Principles**:

- The Red List should be available to all potential users;
- The process of undertaking status assessments of species should be clear and transparent;

- The listings of species should be based on correct use of the categories and criteria and should be open to challenge and correction, based on the categories and criteria, when necessary;
- All status assessments of species should be correctly documented and supported by the best scientific information available;
- The Red List should exist as an electronic version on the World Wide Web to be updated once a year;
- To publish an analysis of the findings of the Red List approximately every five years; and
- The information on the web should be interactive, providing a mechanism to allow people (through appropriate procedures) to provide information for consideration when updating the list.

The *2000 IUCN Red List of Threatened Species* attempts to address the goals, objectives and operating principles outlined above. As a result several innovations have been introduced which will enhance the effectiveness of the Red List as a conservation tool:

- Improved species coverage:
  - All bird species have been completely reassessed by BirdLife International and its partners;
  - All primates have been reassessed following a consultative review workshop on primate systematics;
  - Many other mammals including antelope, bats, cetaceans, otters, wild pigs, wild cattle and wild goats and some rodents were reassessed;
  - Improved coverage of sharks, rays and saw-fish;
  - All Southeast Asian freshwater turtles were comprehensively assessed;
  - A number of new reptile and amphibian assessments from Brazil, the Philippines, Russian Federation and the Russian Republics were carried out;
  - The correction of some insect information and the addition of a number of new European butterfly assessments;
  - Correction of errors in the mollusc listings in the 1996 Red List, a thorough re-evaluation of all the potentially extinct species of mollusc and the inclusion of a number of new assessments;
  - All the trees assessments from *The World List of Threatened Trees* (Oldfield *et al.* 1998) have been incorporated and updated where necessary;
  - All conifers were comprehensively reassessed; and
  - New assessments for plants from Cameroon, Galápagos, Mauritius and South Africa were included as were comprehensive assessments for the carnivorous plant genera *Nepenthes and Sarracenia*, and for the first time almost 100 assessments of mosses have been included.
- Introduction of a Peer review process by the appointment of Red List Authorities responsible for the evaluation of all assessments on the Red List to help ensure the maintenance of standards and the correct application of the criteria (see Annex 1).

- Improved documentation of species on the Red List through (see Annexes 2, 3, 4 and 5):
  - The inclusion of a rationale for many listings explaining how they were reached to improve accountability;
  - Provision of information on range, current population trends, main habitats, major threats and conservation measures taken to gain better insight; into the processes driving extinction; and
  - Improved documentation of extinct species.
- Introduction of a petitions process whereby listings on the Red List can be challenged.
- The Red List information is made more accessible via a new site on the World Wide Web and on CD-ROM:
  - The Web site and CD-ROM provide new innovative and user-friendly ways of interrogating the information on the Red List; and
  - The web site provides a mechanism whereby users can feed corrections and additional information back to the Red List Programme.

The above developments are leading the IUCN/SSC into new technological areas, through which the SSC will be able to provide the information and capacity to perform sophisticated biodiversity analyses. These will contribute significantly to scientific debate and to policies related to conservation at national, regional, and global scales. The Red List Programme through the production of regular, constantly improving and scientifically accurate Red Lists will help ensure that sound, rigorous, and consistent science is used in decision-making and resource-planning.

Analysis

The status of globally threatened species

The conservation status of the world's species is surprisingly poorly known. Only a very small proportion of the world's described species have had their risk of extinction assessed, and there is a strong bias in this sample towards terrestrial vertebrates and plants and in particular to those species found in biologically well-studied parts of the world. Despite these biases, the *IUCN Red List* provides clear evidence that there is indeed cause for concern especially when one considers the marked changes and trends that have become apparent in certain groups in recent years.

This edition of the *IUCN Red List* includes 11,046 species threatened with extinction, 816 species which have already become Extinct or Extinct in the Wild, 4,595 species as Data Deficient or in the Lower Risk categories of conservation dependent and near threatened and a further 1,769 infra-specific taxa or sub-population level assessments. In total, 18,276 taxa are included thereby maintaining the reputation of the *IUCN Red List* as "the most comprehensive list of globally threatened species ever compiled" (Baillie in Baillie and Groombridge 1996, p. Intro 24). This statement is especially true if one also considers the *1997 IUCN Red List of Threatened Plants* (Walter and Gillett 1998) which lists an extraordinary number of 34,000 plants, to be a companion volume.

The 11,046 species threatened with extinction, although less than one per cent of the world's described species, includes 24% of all mammal species and 12% of all bird species. In other words one in every four mammals and one in every eight birds is facing a high risk of extinction in the near future (Table 1). This scale of threat is similar, or possibly even worse for the other vertebrates, especially when one considers the threatened species as a proportion of the rough estimates of those actually evaluated. Using these figures we find that approximately 25% of reptiles, 20% of amphibians and 30% of fishes (mainly freshwater) are listed as threatened (see Table 1 and Figure 1).

For the invertebrate groups, relatively few species have been assessed and the numbers of threatened species (see Table 1) are misleading. Among those invertebrates which have received the most attention there are some groups with large numbers of threatened species, these include 408 primarily inland water crustaceans, 555 insects (mainly butterflies, dragonflies and damselflies), and 938 molluscs (predominantly terrestrial and freshwater species).

For plants, although the total number of 5,611 threatened species sounds very impressive in relation to the numbers for animals, only approximately 4% of described plants have been evaluated using the 1994 IUCN Red List Categories and Criteria. The very large percentage of threatened species (almost 50%) in relation to the total number assessed, is probably an artefact of biases in the data collection. This is especially true when one considers that the *1997 IUCN Red List of Threatened Plants* listed 12.5% of the world's flora as threatened. However, 16% of the conifers (which is the only large, comprehensively assessed plant group in the 2000 Red List), are threatened. So it is possible that the true percentage of plant species that is threatened is much higher than the 1997 results indicated and that the scale of threat is similar to that for some of the animals.

The analysis that follows attempts to explore some of the above results in order to gain a better understanding of which groups are threatened, where the threatened species occur both politically (in countries) and biologically (in biomes and habitats) and what are some of the driving forces causing the extinction crisis. It is hoped that this publication will draw attention to the serious survival crisis now faced by many species world-wide and that the information presented here can help enable decision-makers to take appropriate action before it is too late. The data presented in this Red List provide a rich source of basic information fundamental to understanding the nature and causes of extinction and provides a sound baseline for monitoring extinction trends and the effectiveness of conservation activities. Conservation biologists and scientists are urged to use and explore the information presented in this Red List, as it is only through such open collaboration that we can hope to safeguard the remaining wealth of biological diversity.

The status of animals

For the 1996 Red List, the risk of extinction was evaluated for all known species of birds and mammals. For this edition, once again all the birds were re-evaluated by BirdLife International and its partners (BirdLife International 2000), while for the mammals, the SSC members mainly submitted revised assessments for species on which they had new or better information. The primates were the only mammals to be comprehensively reassessed. Among the other groups of animals, which were only partially assessed in 1996, there has been a slight improvement in the overall coverage, particularly with regards to tortoises and freshwater turtles in south east Asia, the Elasmobranchs (sharks and rays) and the molluscs. New species listings were added in virtually all groups except for the lesser-known invertebrate groups. Animal species that have not been reassessed since 1996 are included in the 2000 Red List with the same listing that they had in 1996. It is the intention of the SSC to reassess and document all these species before 2004.

The 2000 Red List includes 5,435 animal species threatened with extinction compared to 5,205 in 1996 (see Table 1). Although the numbers of threatened species have increased in most groups, none of these increases are significant enough to impact the overall percentages of species threatened in any group. These results were, however, also affected to some extent by an increase in the numbers of described species, especially for mammals, reptiles and fishes, and as a result the percentage of threatened fishes has in fact decreased slightly. Without the increase in numbers of 'new' species, the differences between 1996 and 2000 would have been more marked.

Figure 1. The mammal and bird pie charts represent all the known species for which there is enough information to make a sound conservation assessment. The reptile, amphibian, mollusc, other invertebrates and plant charts represent all assessed species for which there is adequate information to assign a conservation status. The figures used for the number of assessed species are based on figures from WCMC (2000), and corrected using new information from the SSC's threatened species database or from the SSC Specialist Groups. The estimates for the number of species assessed are: reptiles <15%; amphibians <15%; fishes <10%; insects <0.1%; molluscs <5%; crustaceans <5%; other invertebrates <0.1%; plants <4%. CR = Critically Endangered, EN = Endangered, VU = Vulnerable, cd = Lower Risk/conservation dependent, nt = Lower Risk/near threatened, lc and DD = Lower Risk/least concern and Data Deficient.

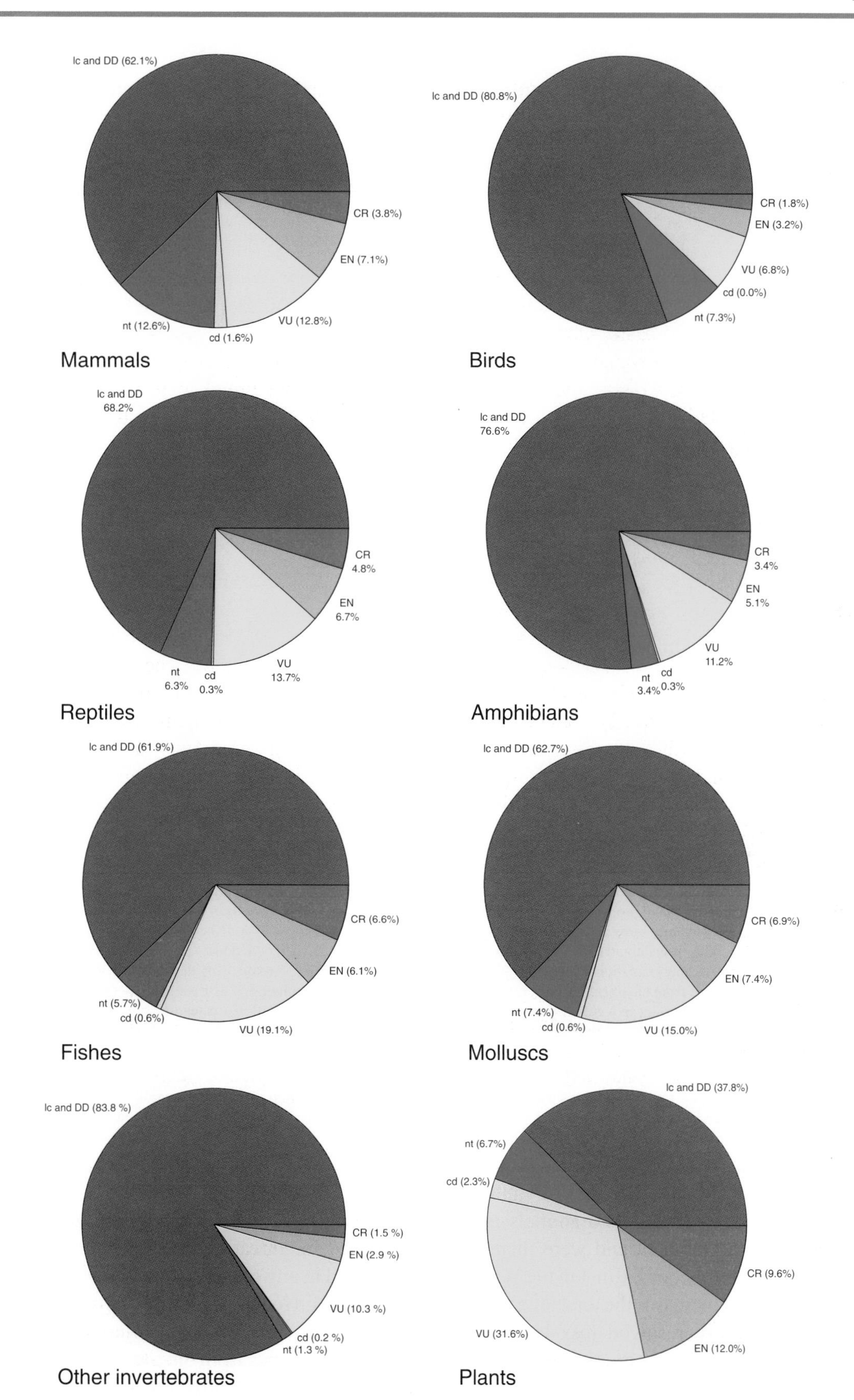

Figure 1. Which species are the most threatened?

Table 1. Numbers of threatened species by major groups of organisms

	Number of species in group	Number of threatened species in 1996	Number of threatened species in 2000	% of total in group threatened in 2000	% of total assessed threatened in 2000*
Vertebrates					
Mammals	4 763	1 096	1 130	24%	24%
Birds	9 946	1 107	1 183	12%	12%
Reptiles	7 970	253	296	4%	25%
Amphibians	4 950	124	146	3%	20%
Fishes	25 000	734	752	3%	30%
Subtotal	51 926	3 314	3 507	7%	19%
Invertebrates					
Insects	950 000	537	555	0.06%	58%
Molluscs	70 000	920	938	1%	27%
Crustaceans	40 000	407	408	1%	20%
Others	130 200	27	27	0.02%	0.2%
Subtotal	1 190 200	1 891	1 928	0.2%	29%
Plants					
Mosses	15 000		80	0.5%	53%
Gymnosperms	876		141	16%	22%
Dicotyledons	194 000		5 099	3%	53%
Monocotyledons	56 000		291	0.5%	26%
Subtotal	265 876		5 611	2%	48%

* **Note:** Threatened includes those listed as Critically Endangered (CR), Endangered (EN) and Vulnerable (VU). Other than mammals and birds only a small or extremely small proportion of the total number of species in any group have been assessed for threatened status. The proportions of species assessed are: birds and mammals 100%; reptiles <15%; amphibians <15%; fishes <10%; insects <0.1%; molluscs <5%; crustaceans <5%; other invertebrates <0.1%; mosses <1%; Gymnosperms 72%; Dicotyledons <5%; and Monocotyledons <4%. These proportions are coarse approximations based on figures from WCMC (2000) and corrected using new information from the SSC's threatened species database or from the SSC Specialist Groups. The plant figures do not include any assessments from the 1997 plants Red List (Walter and Gillett 1998) as these were all done using the pre–1994 IUCN system of threat categorization. Hence the very low proportions of plants assessed compared to the 1997 results. Similarly the results cannot be compared to *The World List of Threatened Trees*, as other plants are now included.

Sources: Species numbers are mostly from WCMC (2000), except for mammals, where we have used and updated the data compiled by Mace and Balmford (2000), and reptiles where we used the recent figures from the EMBL Reptile Database compiled by Peter Uetz (http://www.embl-heidelberg.de/~uetz/LivingReptiles.html), while plants numbers are from Farjon (1998), Hallingbäck and Hodgetts (2000) and Mabberley (1997) as corrected by Rudolph Schmid in *Taxon* 47: 245 (1998).

The total of 5,435 threatened animals is slightly misleading because 1,310 animal species are recorded as Data Deficient and were therefore not assessed for threat. Similarly, the numbers of species listed as Extinct or Extinct in the Wild are not included in any of the threatened totals. Adding the extinct species increases the total number threatened to 6,161. However, as all the extinctions have not been adequately resolved (see later) it is perhaps best to use the lower numbers. For the comprehensively assessed groups, removing the Data Deficient species from the totals and adding the extinct ones has the effect of increasing the percentage of mammals and birds threatened to 27% and

13% respectively. The figure of 23% threatened mammals was considered to be exceptionally high in 1996 and this increase together with that for birds clearly demonstrates that the situation is deteriorating rapidly.

It could be argued that the IUCN criteria are over-listing species and grossly exaggerating the situation. However, comparable information exists for the USA, where The Nature Conservancy (TNC), using different criteria, has carried out a similar comprehensive assessment for 14 of the best-known plant and animal groups representing almost 21,000 species (Master *et al.* 2000). TNC lists 14% of birds and 16% of mammals as threatened. The latter is somewhat lower than the IUCN global estimate, but is still higher than the TNC estimate for the birds. The TNC assessment is probably the most comprehensive national analysis of biodiversity to date, providing complete assessments for groups other than mammals and birds. From this analysis, freshwater-dependent animals such as mussels (69% threatened), crayfishes (51%), stoneflies (43%), fishes (37%) and amphibians (36%) emerge as the most threatened groups in the USA (Master *et al.* 2000). These results clearly reflect the serious vulnerability and degradation of inland water habitats world-wide. The results also indicate that mammals are probably not the most threatened of the world's species as might be assumed from the *IUCN Red List*.

In addition to the species classified as threatened, 1,885 are classified as Lower Risk/near threatened. This category has no specific criteria, and is used for species that come close to meeting the qualifying thresholds for Vulnerable. The vast majority of near threatened animal species are mammals (602 – mainly bats and rodents) and birds (727 – mainly passerines). The other subcategory of Lower Risk is conservation dependent (129 species). This tends to be rarely used for animals except for mammals, which altogether have 74 species in this category, 46 of which are hoofed mammals or artiodactyls.

An examination of the changes within the criteria reveals some marked changes since 1996 (see Table 2). A detailed analysis of these changes is required but it has not been possible to include this here. From the numbers, especially for the comprehensively assessed groups like mammals and birds, there is a clear movement of species from the lower threatened categories into the higher categories. For example, in 1996 there were 168 Critically Endangered birds whereas in 2000 this has increased to 182 species, similarly the figures for Endangered have risen from 235 to 321 species). The reasons for the apparent stability (in mammals) or decrease (in birds) in numbers in the Vulnerable category are not clear. For birds it appears that accurate mapping of all the ranges has resulted in species either being placed into higher categories or that they have wider ranges and so no longer qualify as Vulnerable under Criterion D2 as they did previously (A. Stattersfield, pers. comm.). For the groups not comprehensively assessed, the marked changes are due to a combination of new additions to the list and to changes as a result of reassessments. The changes in the reptiles are almost entirely due to changes in the status of species in the Order Testudines, especially freshwater turtles in Southeast Asia. The number of Critically Endangered Testudines has increased from 10 to 24 and Endangered from 28 to 47 species, reflecting the rapidly deteriorating conservation status of many of these species. Changes to the fishes are largely due to the improved coverage of the Elasmobranchs (sharks and rays). The 1996 Red List included 32 Elasmobranch species, whereas the 2000 edition now has assessments for 95 species, with an increase from seven to 19 species in the Vulnerable category and from seven to 17 in the Endangered category. Although some of these changes are the result of new additions to the Red List, many are genuine changes in status as a result of increasing threats. These trends provide support for the argument that we are rapidly losing biodiversity.

Trends in birds and mammals

The *1996 IUCN Red List* included an analysis of which mammalian and birds orders were most threatened. A comparison of the results with those from the 2000 List shows that these are largely unchanged (Table 3 and see CD-ROM or the web site for further detailed tables). Among the mammals, the largest change was in the number of threatened Primates that have increased from 96 to 116 species. This increase is partly due to a revised taxonomy, but in many cases they are genuine changes brought about by increased habitat loss and hunting. These trends are underlined by the changes in the numbers of primate species within the threatened categories i.e. an increase from 29 to 46 Endangered species and 13 to 19 Critically Endangered species. For the birds, the most significant changes have been in the Procellariformes (albatrosses and petrels) which have increased from 32 to 55 species (all 16 species of albatross are now listed as threatened whereas in 1996 there were only 3—this is due to the impact of longline fisheries) and the Sphenisciformes (penguins) which have doubled in number from five to ten. These increases reflect the increasing threats to the marine environment (BirdLife International 2000). Other bird groups where there have been marked increases in numbers of threatened species include the doves, parrots and passerines (see Table 3), especially among those species occurring in Southeast Asia. This is a direct result of the enormous deforestation that has taken place in places like the Philippines (see BirdLife International 2000).

A number of papers have been published in recent years exploring the 1996 Red List information further to see which orders and families are most susceptible to extinction. For example, a recent paper by Mace and Balmford (2000) examining the mammalian data shows that only the rodents have less threatened species than expected under a binomial distribution, despite them having the largest number of species on the Red List (currently 330); while five orders have significantly more threatened species than would be expected – the Artiodactyla (hoofed animals), Insectivora (shrews), Primates, Perissodactyla (equids, rhinos and tapir) and Sirenia (dugongs and manatees). A significant finding from their analysis is that in general, most of the threatened orders and families of mammals are species-poor and this, coupled with observations from other groups that threatened higher taxa tend to be phylogenetically unique, strongly suggests that impending extinctions will lead to a disproportionate loss of evolutionary novelty (Mace and Balmford 2000). The same non-random extinction trends have also been reported for birds and families like the albatrosses, bustards, cranes, parrots, pheasants and pigeons all have more threatened species than expected (Bennett and Owens 1997; Purvis *et al.* 2000). The bird analyses further show that increased extinction risk is independently associated with increases in body size and low fecundity rates and it is suggested that the evolution of low fecundity many millions of years ago predisposed certain lineage's to extinction (Bennett and Owens 1997).

Trends in reptiles, amphibians and fishes

There has not been a major emphasis on updating the information on reptiles, amphibians and fish for the 2000 Red List, but these species will receive priority attention from the SSC for assessment over the next three years. Nevertheless, the very rapidly deteriorating status of tortoises and freshwater turtles in Southeast Asia has resulted in many important changes in the listings of these species. These species are being heavily exploited for food, and in some cases medicine, and the harvest levels are highly unsustainable, and unregulated. As populations are disappearing in Southeast Asia, there are disturbing signs that the focus of the harvest will shift to the Indian Subcontinent, and perhaps even further afield to the Americas and Africa. It is also known that other Asian species, such as snakes and

Table 2. Changes in numbers of species in the threatened categories (CR, EN, VU) from 1996 to 2000

	CR		EN		VU	
	1996	2000	1996	2000	1996	2000
Group						
Mammals	169	180	315	340	612	610
Birds	168	182	235	321	704	680
Reptiles	41	56	59	74	153	161
Amphibians	18	25	31	38	75	83
Fishes	157	156	134	144	443	452
Insects	44	45	116	118	377	392
Molluscs	257	222	212	237	451	479

Note: Crustaceans and other invertebrates are not included here, as there are virtually no changes in the counts for those groups since 1996.

salamanders, are also the subject of very heavy harvest levels for use in China, but most of these species have not yet been assessed.

It has not been possible in the 2000 Red List to undertake any further analysis on the declining amphibian problem. However, it is already known that a number of amphibian species have been the subject of rapid and unexplained disappearances, for example in Australia, Costa Rica, Panama and Puerto Rico. More attention will be given to this issue in subsequent issues of the Red List. Similarly, relatively little attention has been given to freshwater fish in this Red List, but once again, circumstantial evidence indicates an extremely serious deterioration, especially in the status of riverine species. Master *et al.* (2000) found 37% of freshwater fish species to be threatened in the USA (using the TNC criteria), and it is highly probable that increased attention on these species by SSC over the next three years will confirm a world-wide global crisis in freshwater fish and many other freshwater species.

For marine species, the focus of new information for the 2000 Red List has been on the sharks, rays and skates, where more species have been assessed. It is still too early to draw significant trends from the limited data available, although it is clear that long-lived species with low fecundity are especially at risk, and groups such as the sawfish (*Pristis* sp.) give particular cause for concern.

Trends in invertebrates

Despite the apparently large numbers of threatened invertebrates on the 2000 Red List (1,928 species) this number is proportionally extremely low when one considers that 95% of all known animals are invertebrates (Hammond 1995). The bias towards terrestrial vertebrates in the IUCN Red List is further illustrated by the results of the TNC findings for threatened invertebrates in the USA as described above. The need for a stronger focus on the invertebrate groups has long been recognized, and the SSC is developing a strategy to address this problem (see section on the Red List Programme).

Table 3. Status Category Summary by Major Taxonomic Group (Animals)

Class*	EX	EW	Subtotal	CR	EN	VU	Subtotal	LR/cd	LR/nt	DD	Total
Mammalia	83	4	**87**	180	340	610	**1130**	74	602	240	**2133**
Aves	128	3	**131**	182	321	680	**1183**	3	727	79	**2123**
Reptilia	21	1	**22**	56	79	161	**296**	3	74	59	**454**
Amphibia	5	0	**5**	25	38	83	**146**	2	25	53	**231**
Cephalaspidomorphi	1	0	**1**	0	1	2	**3**	0	5	3	**12**
Elasmobranchii	0	0	**0**	3	17	19	**39**	4	35	17	**95**
Actinopterygii	80	11	**91**	152	126	431	**709**	12	96	251	**1159**
Sarcopterygii	0	0	**0**	1	0	0	**1**	0	0	0	**1**
Echinoidea	0	0	**0**	0	0	0	**0**	0	1	0	**1**
Arachnida	0	0	**0**	0	1	9	**10**	0	1	7	**18**
Chilopoda	0	0	**0**	0	0	1	**1**	0	0	0	**1**
Crustacea	8	1	**9**	56	72	280	**408**	9	1	32	**459**
Insecta	72	1	**73**	45	118	392	**555**	3	76	40	**747**
Merostomata	0	0	**0**	0	0	0	**0**	0	1	3	**4**
Onychophora	3	0	**3**	1	3	2	**6**	0	1	1	**11**
Hirudinoidea	0	0	**0**	0	0	0	**0**	0	1	0	**1**
Oligochaeta	0	0	**0**	1	0	4	**5**	0	1	0	**6**
Polychaeta	0	0	**0**	1	0	0	**1**	0	0	1	**2**
Bivalvia	31	0	**31**	52	28	12	**92**	5	60	7	**195**
Gastropoda	260	12	**272**	170	209	467	**846**	14	177	513	**1822**
Enopla	0	0	**0**	0	0	2	**2**	0	1	3	**6**
Turbellaria	1	0	**1**	0	0	0	**0**	0	0	0	**1**
Anthozoa	0	0	**0**	0	0	2	**2**	0	0	1	**3**
Total	693	33	**726**	925	1353	3157	**5435**	129	1885	1310	**9485**

*Mammalia (mammals), Aves (birds), Reptilia (reptiles), Amphibia (amphibians), Cephalaspidomorphi (lampreys and hag fish), Elasmobranchii (sharks, skates, rays and chimaeras), Actinopterygii (bony fishes), Sarcopterygii (coelacanth), Echinoidea (sea urchins, starfish, etc), Arachnida (spiders and scorpions), Chilopoda (centipedes), Crustacea (crustaceans), Insecta (insects), Merostomata (horseshoe crabs), Onychopora (velvet worms), Hirudinoidea (leeches), Oligochaeta (earthworms), Polychaeta (marine bristle worms), Bivalvia (mussels and clams), Gastropoda (snails, etc.), Enopla (nemertine worms), Turbellaria (flatworms), Anthozoa (sea anemones and corals). EX – Extinct, EW – Extinct in the Wild, CR – Critically Endangered, EN – Endangered, VU – Vulnerable, LR/cd – Lower Risk/ conservation dependent, LR/nt – Lower Risk/near threatened, DD – Data Deficient.

In the interim, the list of invertebrates is slowly increasing, with the most significant changes taking place among the molluscs. Less than five per cent of molluscs have been assessed and these assessments have largely been confined to the terrestrial and freshwater species. The number of land-snails world-wide is estimated to eventually reach between 25,000 and 35,000 species, and so far only a very small proportion of the species has been assessed. The majority of the assessments relate to the better known regions such as the USA, Europe, Australia as well as the recognized areas of endemism on islands.

There are approximately 5,000 species of freshwater molluscs world-wide, of which about 1,000 are bivalves and 4,000 are gastropods. This species inventory is far from complete and new species are still regularly being described from all parts of the world, including from better known regions like Europe, USA, Japan and Australia. Despite incomplete inventorying, global patterns of hotspots in terms of diversity and endemism clearly emerge. There are also three main types of freshwater habitat that are particularly critical to freshwater mollusc conservation, which also have a very diverse fauna and are highly vulnerable: rivers, springs and underground aquifers, and ancient oligotrophic lakes. The spring snails represent 19% of all threatened molluscs, which is a major increase over the 12% listed in 1996. While there are a number of genuine increases in the numbers of threatened molluscs (see the Vulnerable and Endangered categories in Table 2), some of the more marked changes such as the decrease from 257 to 222 Critically Endangered species is reflective of better knowledge (Table 2). A large number of species were suspected to be extinct, but were not listed as such in 1996 because of the lack of adequate survey information and as a result were placed under the Critically Endangered category. New survey information now confirms the extinction of these species, an event, which probably took place during the 1930s when rivers and springs were heavily impacted in the eastern USA. Other changes in numbers are also a result of the clarification of taxonomy and the removal of a number of duplicate entries from the Red List (these were mainly in the lower categories, especially Data Deficient).

The status of plants

A major difference between the 1996 and the 2000 Red Lists is the inclusion of plant species, which have been assessed using the 1994 IUCN Red List Criteria. This has more than doubled the number of species on the Red List, as all 7,388 species (includes species in all categories from Data Deficient to Extinct) listed in *The World List of Threatened Trees* (Oldfield *et al.* 1998) have been included. As these tree assessments were all done relatively recently there was no need for any reassessments, but some changes were made in the light of new information and the SSC Conifer Specialist Group in their preparation of an Action Plan, reassessed many of the conifers (Farjon and Page 1999). In addition to the trees more than 300 new assessments encompassing a wide range of additional trees and other plants from Cameroon, Galápagos, Mauritius and South Africa, many of which were not listed in the 1997 plants Red List, were included. The additions also included comprehensive assessments for the carnivorous plant genera *Nepenthes and Sarracenia*, and approximately 100 bryophytes (mosses and liverworts), the first time this group has ever appeared in an *IUCN Red List*.

As with the invertebrates, the seemingly very large figure of 5,611 threatened plant species is proportionally very small relative to the total number of plant species world-wide (see Table 1). It is therefore premature at this stage to attempt any detailed analysis of the plants as the low numbers assessed and the strong bias towards trees and certain geographic areas misrepresents the overall picture for plants. For further details on the numbers of plants in each category, readers are referred to Tables 1, 3 and 4 and to the detailed order and family results on the CD-ROM or web site. As stated previously, the only group to be comprehensively assessed, are the conifers, and although 16% of these are threatened, any extrapolation of these trends to other species could be misleading. However, this high proportion of threatened species is confirmed by the results of the TNC analysis (Master *et al.* 2000) which indicates 24% of USA conifers to be threatened. Similarly the *1997 IUCN Red List of Threatened Plants* (Walter and Gillett 1998) lists 30% of conifers in the old Endangered (E) and Vulnerable (V) categories. As with many of the threatened vertebrates, the conifers are an ancient lineage indicating that perhaps non-random extinction is also happening in the plant Kingdom. A

Table 4. Status Category Summary by Major Taxonomic Group (Plants)

Class*	EX	EW	Subtotal	CR	EN	VU	Subtotal	LR/cd	LR/nt	DD	Total
Bryopsida	2	0	**2**	10	15	11	**36**	0	0	0	**38**
Anthocerotopsida	0	0	**0**	0	1	1	**2**	0	0	0	**2**
Marchantiopsida	1	0	**1**	12	16	14	**42**	0	0	0	**43**
Coniferopsida	0	1	**1**	17	40	83	**140**	24	52	33	**250**
Ginkgoopsida	0	0	**0**	0	1	0	**1**	0	0	0	**1**
Magnoliopsida	69	14	**83**	896	1110	3093	**5099**	203	610	298	**6293**
Liliopsida	1	2	**3**	79	83	129	**291**	17	45	39	**395**
Subtotal	73	17	**90**	1014	1266	3331	**5611**	244	707	370	**7022**

* Bryopsida (mosses), Anthocerotopsida (hornworts), Marchantiopsida (liverworts), Coniferopsida (conifers), Ginkgoopsida (ginkgo), Magnoliopsida (dicotyledons), Liliopsida (monocotyledons). EX – Extinct, EW – Extinct in the Wild, CR – Critically Endangered, EN – Endangered, VU – Vulnerable, LR/cd – Lower Risk/conservation dependent, LR/nt – Lower Risk/near threatened, DD – Data Deficient.

further very sobering result from the TNC work is that one third (33%) of the 15,300 native flowering plants in the USA are threatened with extinction (Master *et al.* 2000). This number of threatened species is more than one order of magnitude larger than that of any other group assessed in the USA. The SSC has a very ambitious programme to increase the coverage of plants on the *IUCN Red List* over the next few years and it will be important to determine whether or not the USA trends are repeated elsewhere in the world.

Where are the threatened species?

Countries with the largest numbers of threatened species

An analysis and identification of the countries with the largest number of threatened species enables countries to be informed about their global responsibility to protect and maintain the biodiversity for which they are ultimately the stewards. An analysis of this type, however, is fraught with several difficulties:

- Depending on the group being examined, but particularly for those which are not comprehensively assessed, there are geographic biases as some countries such as Australia, South Africa and the USA, have been subjected to detailed review, and may therefore have proportionally more species listed than countries in less well studied regions.
- There are also countries where an enormous amount of work on threatened species has been done, but the IUCN Red List Category system was not used, making it difficult to incorporate the results into the *IUCN Red List*.
- Countries with large surface areas are more likely to dominate a 'top 20' list.
- The alternative of using the percentage of species that are threatened compared to overall diversity also creates problems because:

- Countries with fewer species are likely to create greater error rates than countries with many species. For example, New Zealand has only four terrestrial mammals of which three are threatened. If two were incorrectly assessed, then instead of 75% of mammals being threatened there would only be 25%. In contrast, Indonesia has 561 terrestrial mammals, 135 of which are threatened. An incorrect assessment of two threatened species would only change the percentage threatened from 24.1 to 23.7%.
- Countries with small sample sizes also pose statistical problems as they have very few discrete values and the data may not be usable for analyses that require the dependent variable to be continuous.

Although the usefulness of measuring the number and proportion of threatened species in a country is limited in statistical analyses, when combined they can provide a useful indication of high diversity areas that have disproportionately more threatened species. This section focuses primarily on mammals and birds, as they are the only two groups to have been comprehensively assessed. The results for plants are also included based on the assumption that the large numbers of trees assessed globally may indicate some patterns that are different to those for the mammals or birds. For the birds and mammals, the total numbers of species in each country were derived from the recent book on global biodiversity produced by the World Conservation Monitoring Centre (WCMC 2000). In using these figures, it was assumed that any taxonomic differences with the Red List would be negligible. However, given the recent significant changes in primate taxonomy, which have been adopted in the 2000 Red List, this may affect some results slightly. A further problem with the mammal data is that the distributions of many of the cetaceans are not recorded by country, hence these species were excluded from the analysis. The bird data also presented a problem in that the total numbers used were for breeding birds only (WCMC 2000). The numbers of threatened birds recorded per country in the Red List, however, include both breeding and non-breeding species. The only other data set available with numbers of birds per country included every possible vagrant thereby vastly inflating the totals for each country. The WCMC data was used for the analysis, but caution must be exercised in using the results.

Mammals

The results for the mammals are shown in Figures 2a, b and c. Figure 2a indicates that as in 1996, Indonesia (135 species) still has by far the highest number of threatened mammals of any country in the world. India (80 species) and Brazil (75 species) have now both moved ahead of China (72 species). While lower down the list there are some major changes with Thailand (32 species) and the USA (29 species) being displaced from the top twenty by Cameroon (38 species) and the Russian Federation (35 species). Tanzania (38 species) has also moved up from 20th to 14th position. Comparing the results in Figures 2a, b and c shows that countries with high numbers of threatened species are not necessarily those with the highest percentages of their mammal faunas that are threatened. In particular, although seven African countries feature in the top 20 of total numbers of threatened species, only one of these, Madagascar (48 species or 34%) is within the upper quartile for the percentage of mammal species threatened. The closest match in countries with both high numbers of threatened species and high percentages of total mammalian fauna are in Southeast Asia and Australasia, which is a similar result to that described in the 1996 Red List.

If the number of threatened species is plotted against the total mammal diversity for each country and a regression analysis is performed, the resulting scatter plot shows a number of countries are situated well above the regression line. These are countries which have more threatened species than

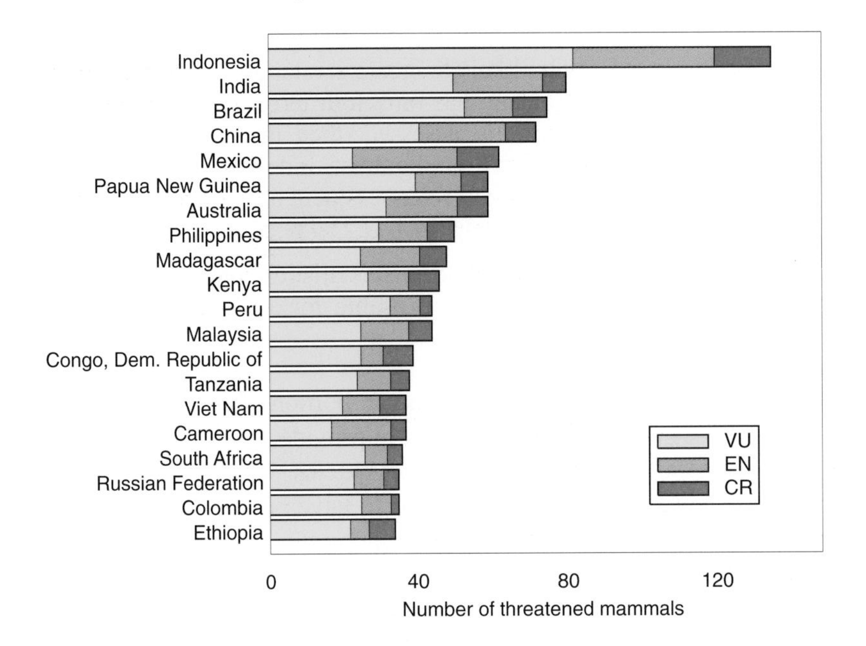

Figure 2a. The twenty countries with the largest numbers of threatened mammal species ranked from the highest to the lowest (the colour coding of the histogram illustrates the relative contributions of the three threatened categories).

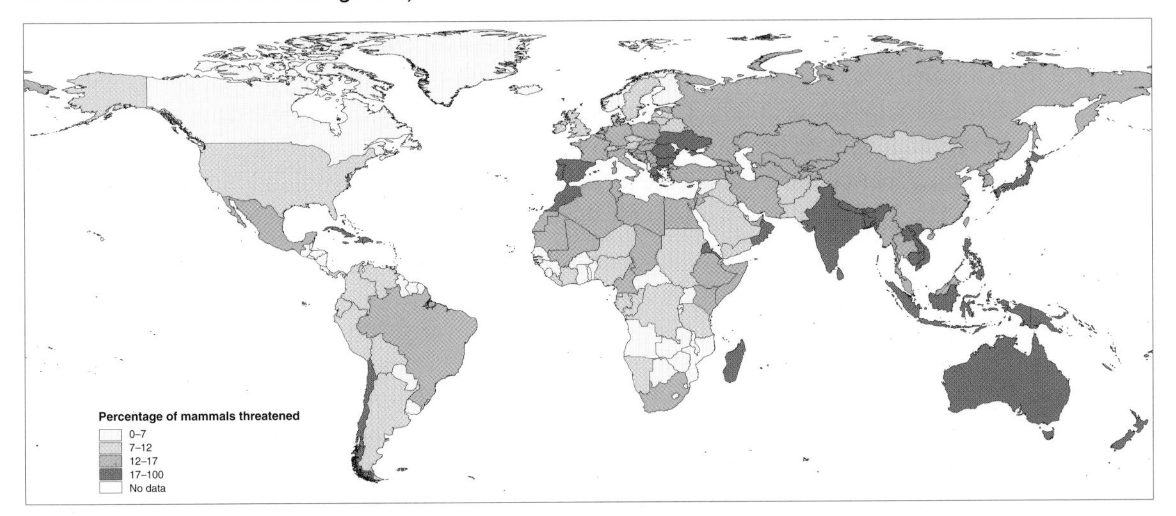

Figure 2b. The countries with the highest percentage of threatened mammals relative to the total number of mammals recorded for each country (mammal totals from WCMC 2000). The four categories on the map are quartile splits. For example, the red colour represents the top 25% of countries that have the highest proportion of threatened species. In each quartile there are approximately fifty countries.

would be expected, and the 25 countries for which the number of threatened species is most in excess of what is expected are shown in Table 5. Nineteen of these countries are island states (including Australia), which provides support for the biogeographical analysis of Mace and Balmford (2000) which indicated that species restricted to islands, wherever they are, have a higher level of threat than continental species in the same biogeographic region, and that island species are generally more vulnerable to extinction. The inclusion of mainland countries like India, Brazil, China, Bhutan and Viet Nam amongst all these islands is an indication of particularly serious threats in these countries.

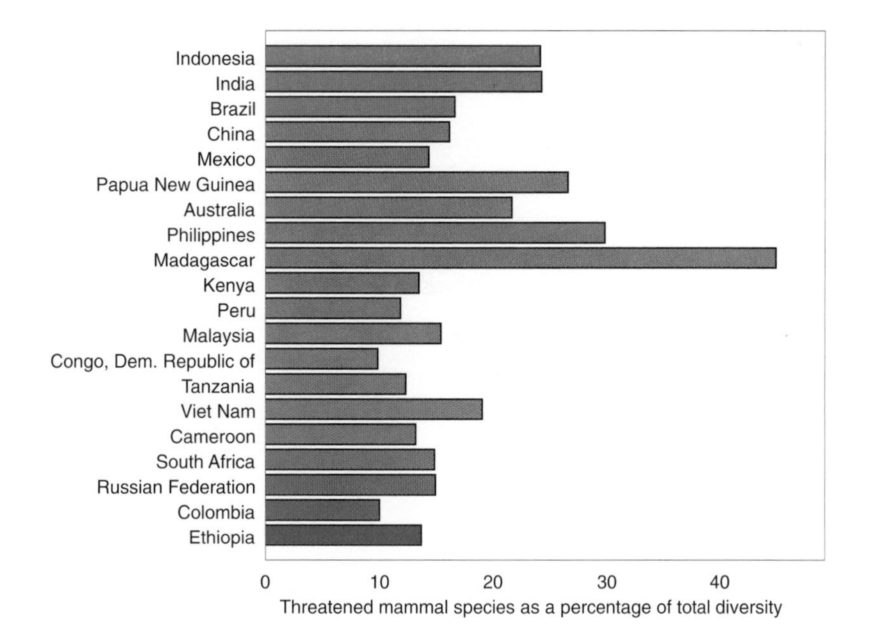

Figure 2c. The twenty countries with the largest numbers of threatened mammal species, with their numbers as a percentage of total mammal diversity found in each country (mammal totals from WCMC 2000).

Birds

The results for the birds are depicted in the same manner as for the mammals (see above) in Figures 3a, b and c. Indonesia emerges here once again as the country with the most threatened birds, followed closely by Brazil with 115 and 113 species respectively. Then follow Colombia, China, Peru and India with 78, 76, 75 and 74 species respectively. The overall results are very similar to those for 1996 with all the same countries appearing except for Papua New Guinea (32 species) which is now replaced by Tanzania (33 species). The ranking of the countries has changed because of increases in numbers of threatened species. The countries with by far the highest percentage of threatened species are New Zealand and the Philippines with 42% and 35% respectively, which matches the findings in 1996. However, the values this time are considerably larger due in part to an increase in the number of threatened species in New Zealand (from 44 to 62), but primarily due to the fact that the total number of birds threatened are being compared only to the breeding birds. In 1996 the total numbers of birds for New Zealand and Philippines were given as 287 and 556 respectively, whereas the breeding figures are only 150 and 196 (WCMC 2000). In 1996, the results indicated that the New World and Asia were the key areas to consider, while the results for 2000 suggest that North America and parts of Africa should also be considered (Figure 3b).

Using maps in Figures 2b and 3b for conservation planning has limited utility. They can indicate key countries at a global level, but often these countries like Australia or Brazil are so large that one needs to have additional information to start any conservation planning. BirdLife International has mapped the distributions of all threatened birds and has compiled these into a single density distribution map (Figure 4), which is a far more accurate tool for conservation planning than the map in Figure 3 as it is not constrained by political boundaries. This shows that globally threatened birds are unevenly distributed. They occur on more than 20% of the earth's land surface but less than 5%

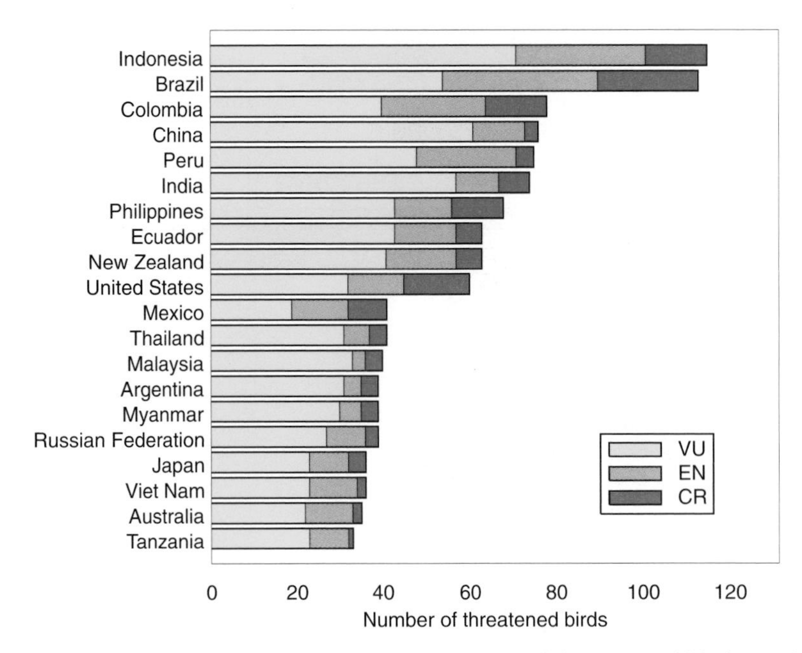

Figure 3a. The twenty countries with the largest numbers of threatened bird species ranked from the highest to the lowest.

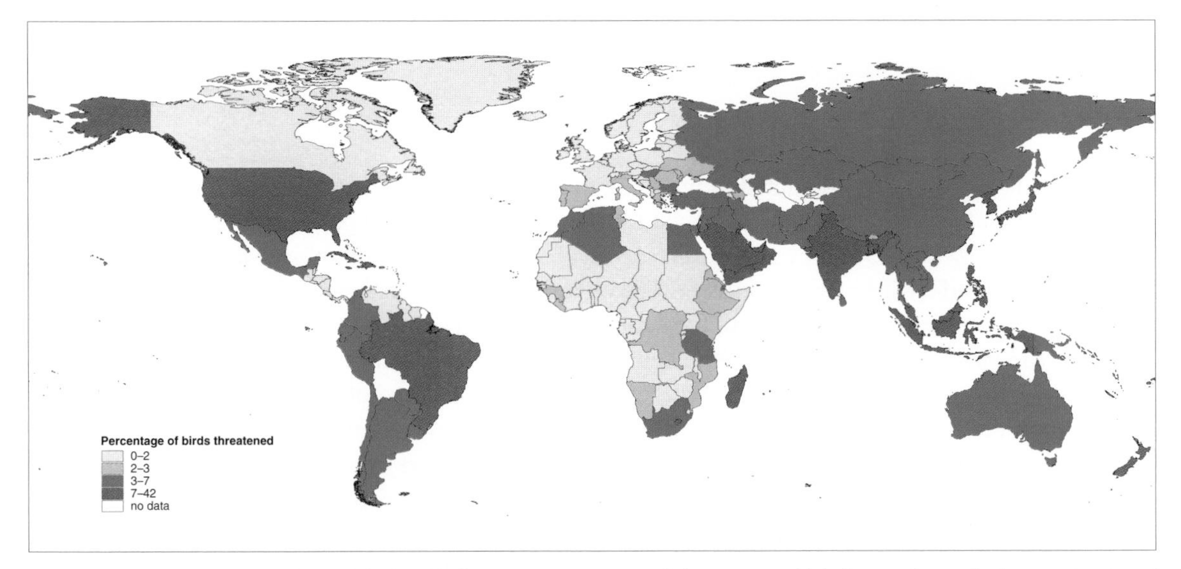

Figure 3b. The countries with the highest percentage of threatened bird species relative to the total number of breeding birds recorded for each country (breeding bird totals from WCMC 2000).

holds almost 75% of all threatened birds (BirdLife International 2000). This allows conservation planners to target resources in those areas where the extinction risk to birds, and therefore often the threat to the wider environment, is greatest. For example, in Brazil while it is evident that much of Amazonia is important, it is the coastal region where the Atlantic forest occurs which is of prime conservation importance and this applies equally to both birds and mammals, especially the primates (Rylands *et al.* 1996).

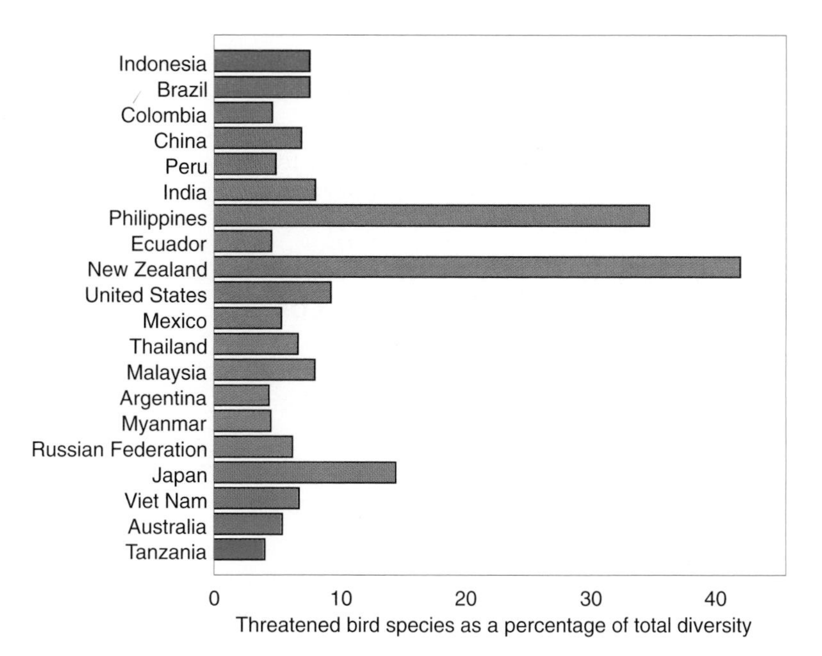

Figure 3c. The twenty countries with the largest numbers of threatened bird species, with their numbers as a percentage of total breeding bird diversity recorded for each country (totals of breeding birds from WCMC 2000).

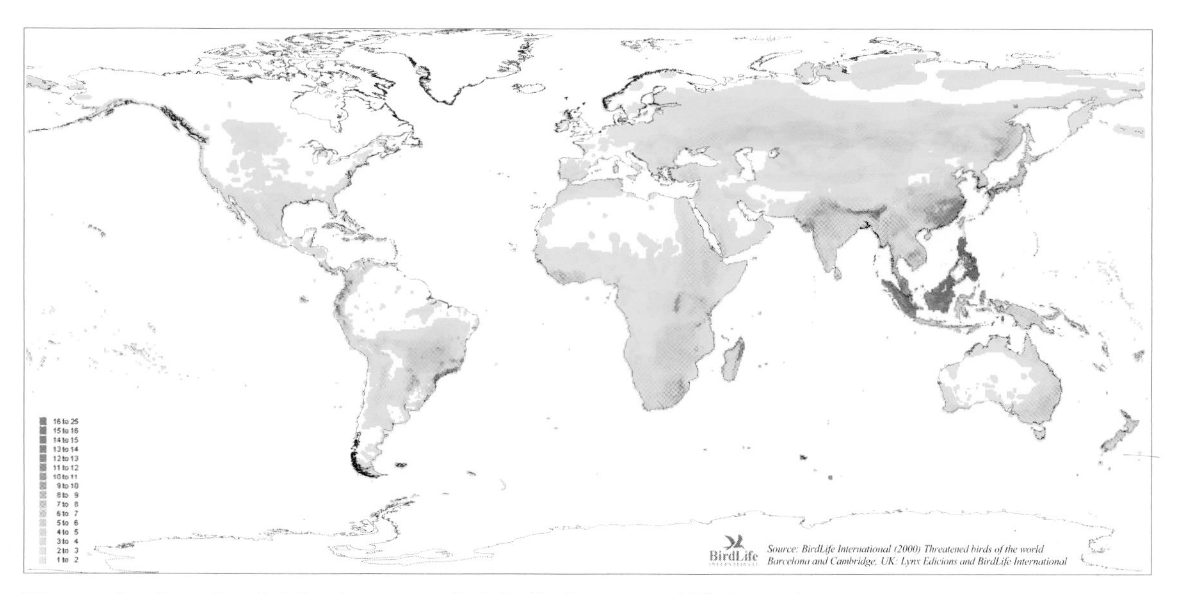

Figure 4. Density distribution map of globally threatened bird species.

Figure 4 confirms that threatened birds are concentrated in the Neotropics and Southeast Asia, reflecting the relative species-richness and higher levels of local endemism in these regions. Brazil (113 threatened species), Indonesia (115) and Colombia (78) have the highest totals, but the political responsibility for conserving threatened species is shared since 219 territories have at least one threatened bird (BirdLife International 2000).

The countries with more than expected numbers of threatened birds (Table 5) supports the results from the mammals, although at first glance it appears to comprise fewer island states and more continental areas than the mammals. This is probably not the case because the US figures are probably dominated by birds from Hawaii rather than the mainland, for Malaysia the birds are possibly from Sabah and/or Sarawak while Ecuador probably refers to Galápagos.

Table 5. Countries and territories with more threatened species than expected (arranged in descending order)

Rank	Mammals	Birds
1	Madagascar	New Zealand
2	Indonesia	Philippines
3	Micronesia	Indonesia
4	India	Brazil
5	Philippines	Western Samoa
6	Papua New Guinea	Madagascar
7	Solomon Islands	Fiji
8	Seychelles	Cook Islands
9	Mauritius	India
10	Guam (to USA)	Solomon Islands
11	Australia	China
12	Japan	Vanuatu
13	New Zealand	United States
14	Palau	South Korea
15	Fiji	São Tomé and Principe
16	New Caledonia (to France)	Northern Mariana Islands (to USA)
17	Northern Mariana Islands (to USA)	Mauritius
18	Brazil	Japan
19	China	Colombia
20	Guadeloupe (to France)	Russian Federation
21	Réunion (to France)	Malaysia
22	Western Samoa	Peru
23	Bhutan	New Caledonia (to France)
24	Viet Nam	Comoros
25	Bangladesh	Ecuador

Plants

The distribution of the threatened plants (primarily trees) and the top 20 countries with the most threatened plants are shown in Figures 5a and b respectively. Although the number of plants assessed represent only a very small proportion of those actually threatened, these results indicate that the northern temperate regions are not important in terms of threatened species, and that it is primarily the tropical regions in South and Central America, Central and West Africa and Southeast Asia that emerge as being the most important areas. To some extent this is corroborated by results in the *Centres of Plant Diversity* series (WWF and IUCN 1994). However, there are a number of important plant diversity areas, which are known to have high numbers of threatened species e.g., Australia and South

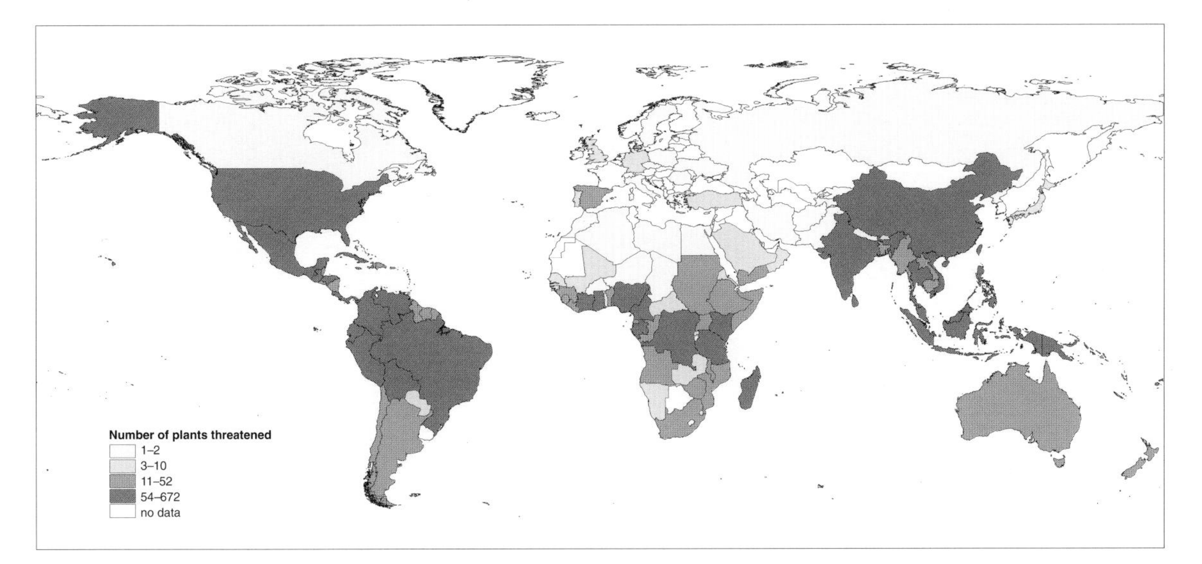

Figure 5a. The countries with the highest numbers of threatened plant species.

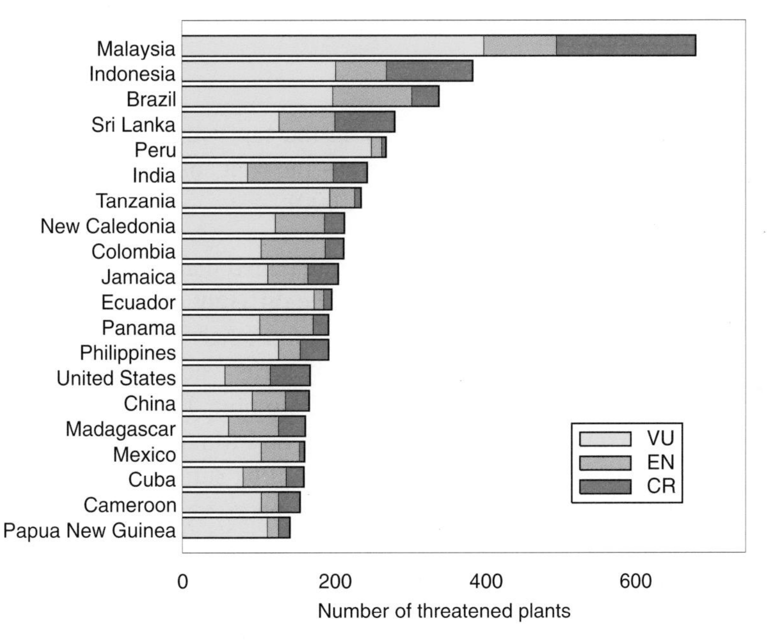

Figure 5b. The twenty countries with the largest numbers of threatened plant species ranked from the highest to the lowest.

Africa, which do not feature in relation to the tropics. This is primarily due to biases in the data set. Very few plant species in either of these countries have been assessed using the 1994 IUCN Red List Categories.

In terms of the top 20 countries, Malaysia has by far the most threatened species with an extremely high total of 681 species, of which a large proportion are dipterocarp trees which are highly sought after by logging companies. This is followed by Indonesia, Brazil and Sri Lanka with 384, 338, and 280 threatened species respectively. As none of the floras of any of these countries have been comprehensively assessed it is not realistic to present this information as percentages of the total flora.

Figure 6. Congruency between the top twenty countries with the largest numbers of threatened species of mammals, birds and plants.

The USA is the only largely non-tropical country in the top twenty. However, the Red List only includes 168 threatened species for the USA whereas the TNC assessments suggest at least 5,090 flowering plants alone are threatened (Master *et al.* 2000).

Combining the results from the total numbers of threatened mammals, birds and plants provides some indication of the degree of congruence between the different groups (see Figure 6). These very preliminary results indicate that there is not 100% congruence in any of these groups and that conservation planning has to bear this in mind. The only areas where conservation will clearly be of benefit to all three groups is in the Neotropics (Brazil, Colombia, Ecuador and Mexico), East Africa (Tanzania) and Southeast Asia (China, India, Indonesia and Malaysia). For mammal conservation Africa emerges as important, while for birds alone, Argentina surprisingly and the Southeast Asian block of Myanmar, Viet Nam, Cambodia and Lao PDR emerge as being important.

Reptiles, amphibians, fishes, invertebrates and molluscs

The taxonomic groups shown in Figure 7 have not been comprehensively assessed and are therefore subject to regional biases based on information availability. For example, the USA features as the country with the most threatened species among fish (for all fish species and for freshwater fish alone) and invertebrates (both excluding molluscs and for molluscs alone). It is highly unlikely that the USA has more than twice as many globally threatened invertebrates as any other country. The reason is simply that the status of inland water crustaceans and certain insect groups is particularly well known in the USA (see Master *et al.* 2000). Similarly, the histogram for the molluscs is a classic guide to where malacologists are based or have been working and is a tribute to the excellent work done by many SSC Mollusc Specialist Group (MSG) members.

Although the sequences of countries and numbers of threatened species are very incomplete, Figure 7 identifies a number of countries with significantly high numbers of threatened species, especially in the reptiles and amphibians. The appearance of ten Southeast Asian countries in the top twenty as opposed to six in 1996, is largely a result of the reassessment of all freshwater turtles in the region at a workshop held in Cambodia last year by the Asian Turtle Trade Working Group. These changes reflect a genuine change in the region due to increased demand for animals for the food trade. The appearance of countries such as Turkey, Greece and Puerto Rico in the figures for reptiles, amphibians and fishes suggests that these groups might display very different patterns to those shown by mammals and birds, once a comprehensive assessment of their status has been completed. The detailed country tables for all these taxa are included on the web site and the CD-ROM should the reader wish to explore them further.

Table 6. Numbers of threatened species in three major biomes

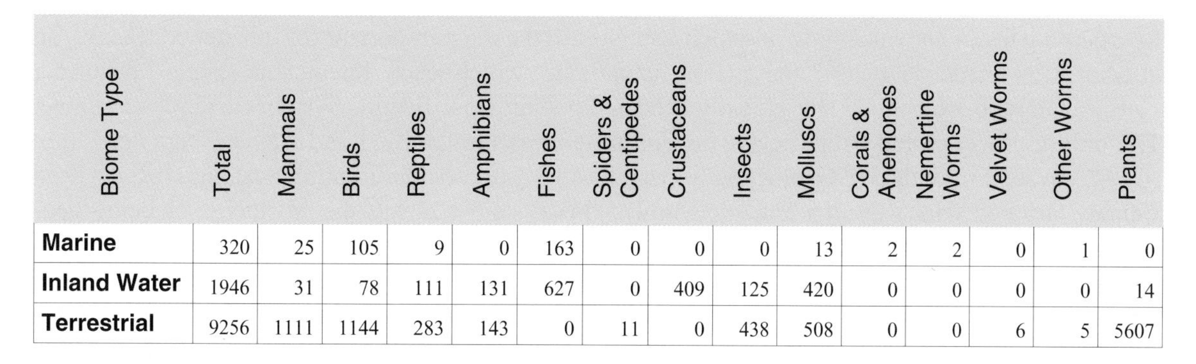

Biome Type	Total	Mammals	Birds	Reptiles	Amphibians	Fishes	Spiders & Centipedes	Crustaceans	Insects	Molluscs	Corals & Anemones	Nemertine Worms	Velvet Worms	Other Worms	Plants
Marine	320	25	105	9	0	163	0	0	0	13	2	2	0	1	0
Inland Water	1946	31	78	111	131	627	0	409	125	420	0	0	0	0	14
Terrestrial	9256	1111	1144	283	143	0	11	0	438	508	0	0	6	5	5607

Note: The counts are only for globally threatened species (CR, EN, VU). All species have been assigned to one or more biome, for example, seals are both marine and terrestrial, otters and amphibians are inland water and terrestrial, diadramous fishes (e.g. sturgeon) are inland water and marine, etc. The molluscs were assigned with the help of Mary Seddon of the SSC Mollusc Specialist Group. Inland waters include saline water bodies, cave waters, freshwaters, etc. The plants, because they largely comprise trees, have been classified mainly as terrestrial, but some of the mosses included grow partially submerged in freshwater streams.

Distribution of threatened species by biomes

From the taxonomic groups included on the Red List it has been assumed that the preponderance of species would be terrestrial, especially since very little is known about the population status of most aquatic organisms. An attempt has been made to code the presence of all threatened species according to their occurrence in the three major biomes – terrestrial, inland waters and marine (Table 6). Where large numbers of aquatic species have been assessed, the proportion of threatened species rises markedly. The steep increase in the number of threatened marine birds compared to the figure of 61 given by WCMC (2000) is attributed to the increased number of albatrosses, petrels and penguins now included in the threatened categories. The low number of marine molluscs is partly a reflection of the lack of data on marine species, but is also due to the fact that most marine molluscs tend to be widespread (M. Seddon pers. comm.). However, some widespread marine clams are in serious trouble due to over-exploitation impacting reproductive output and resulting in the lack of recruitment (depensation effects). The lack of marine fishes on the Red List has largely been due to a lack of knowledge, but this situation is slowly changing with increasing attention being focussed on many range restricted coral reef fish, the groupers and wrasses and the sharks and rays. The sharks and rays including the chimaeras, have been targeted for complete assessment by 2003. The number of marine fishes assessed and listed as threatened is expected to rise sharply over the next few years. The number of threatened inland water species has increased in all groups except for the molluscs compared to the figures given in WCMC (2000). A large proportion of these species are found in the USA which has an extremely rich freshwater biota comprising 61% of the world's crayfishes, 29% of freshwater mussels, 17% of freshwater snails and 10% of freshwater fishes (Stein *et al.* 2000). As shown previously, a large proportion of these are considered to be threatened, (for example 69% of freshwater mussels (Master *et al.* 2000)), and the change in the total number of molluscs from the figure given by WCMC is due to the acceptance that many of these species are in fact now Extinct. These results are a clear indication of the extremely vulnerable nature of freshwater habitats and species occurring in these systems are likely to be facing a much higher risk of extinction than their counterparts in the terrestrial and marine environments.

Distribution of threatened species by major habitats

In order to gain an understanding of which habitats are the most important for threatened species, an attempt was made to record the major habitats in which each threatened species occurred. Unfortunately, there is no single globally accepted habitat classification system currently available. The only uniform global habitat classification scheme is the Global Land Cover Characterization (GLCC) system from the US Geological Service Earth Resources Observation Systems (EROS) Data Center (http://edcdaac.usgs.gov/glcc/glcc.html). This system also has the advantage of being geo-spatially explicit. A simplified version of this scheme was developed in which the habitat categories (shown in Annex 4) are the result of an amalgamation of many of the categories within the GLCC system. Some modifications to the system were required to accommodate some of the inland water and marine habitats.

BirdLife International has used its own system in assigning habitat types to 1,180 of the threatened bird species (almost 100%). Fortunately, the BirdLife system is relatively similar to the simplified GLCC system so most of the categories could be converted to the GLCC habitat types. For the mammals, the GLCC habitats were assigned to species based on the documentation provided with new submissions, and based on surveys of the available literature. Habitat types were successfully assigned to 515 (46%) of the threatened mammal species. Unfortunately the habitat data recorded for the threatened trees in *The World List of Threatened Trees* did not lend itself to such conversion. The histograms in Figure 8 show the top 20 habitat types for birds and mammals. It is important to note that the results within each histogram are non-exclusive as the same species can be assigned to more than one habitat type.

Comparing the top six most important habitats for threatened birds and mammals, five of these habitats are common to both although the rankings between them vary only slightly. There is complete agreement on the top two habitats for both groups, namely lowland and montane tropical rainforest. The analysis of the bird habitats by BirdLife (BirdLife International 2000) indicates that threatened birds are highly habitat-restricted with 883 species (74%) almost entirely dependent on a single habitat type. Of these, 75% are dependent on forests. More than 900 threatened bird species use tropical rainforests and 42% of these are found in lowland rain forest while 35% occur in montane rain

Figure 7. The twenty countries with the largest numbers of threatened species for reptiles; amphibians; all fish (freshwater and marine); freshwater fish only; invertebrates excluding molluscs; and molluscs. The results shown here are strongly affected by sampling bias, resulting in countries such as the USA and Australia scoring much higher in relative terms than they will once sampling has been completed.

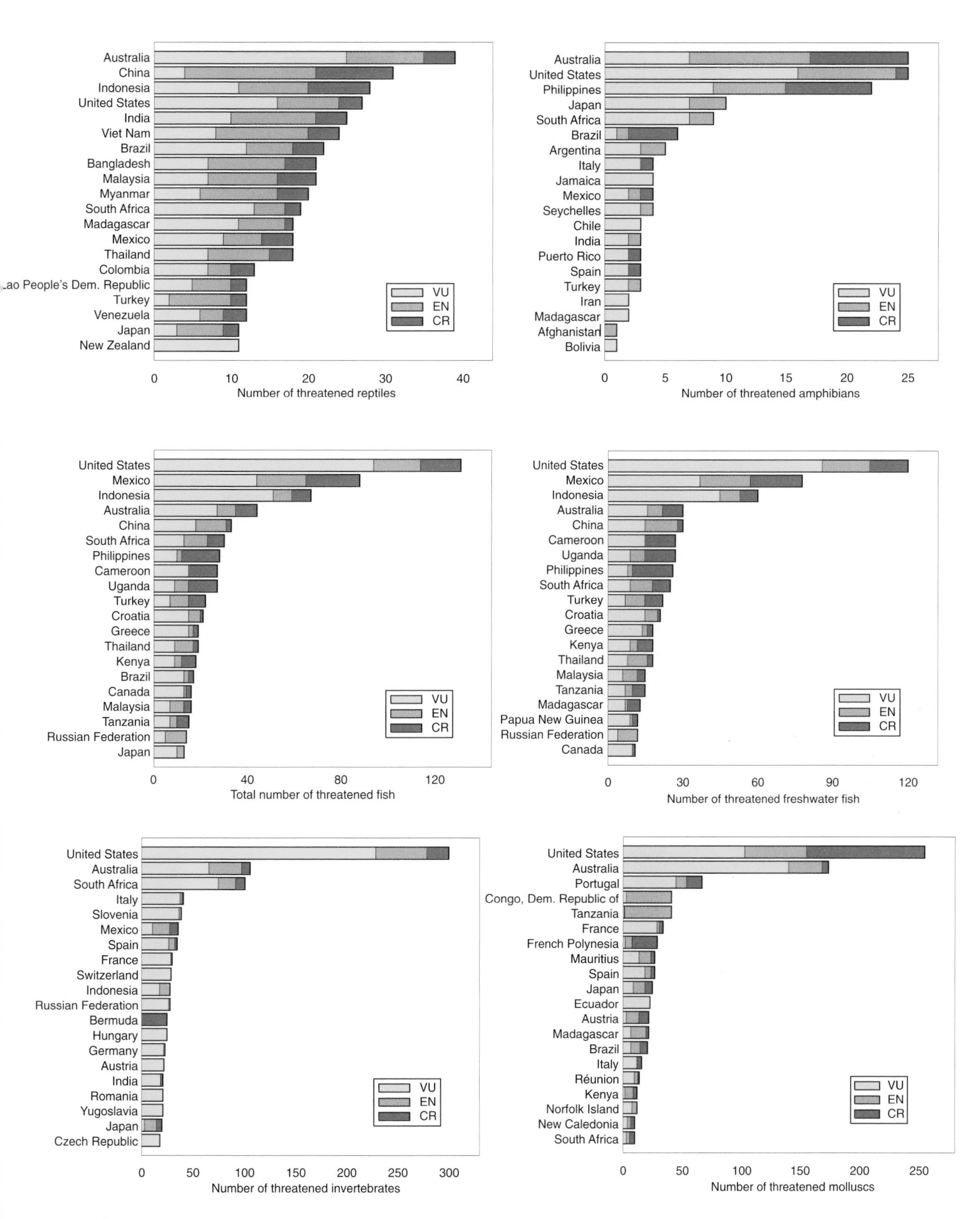

Figure 7. Where are the other threatened animal species?

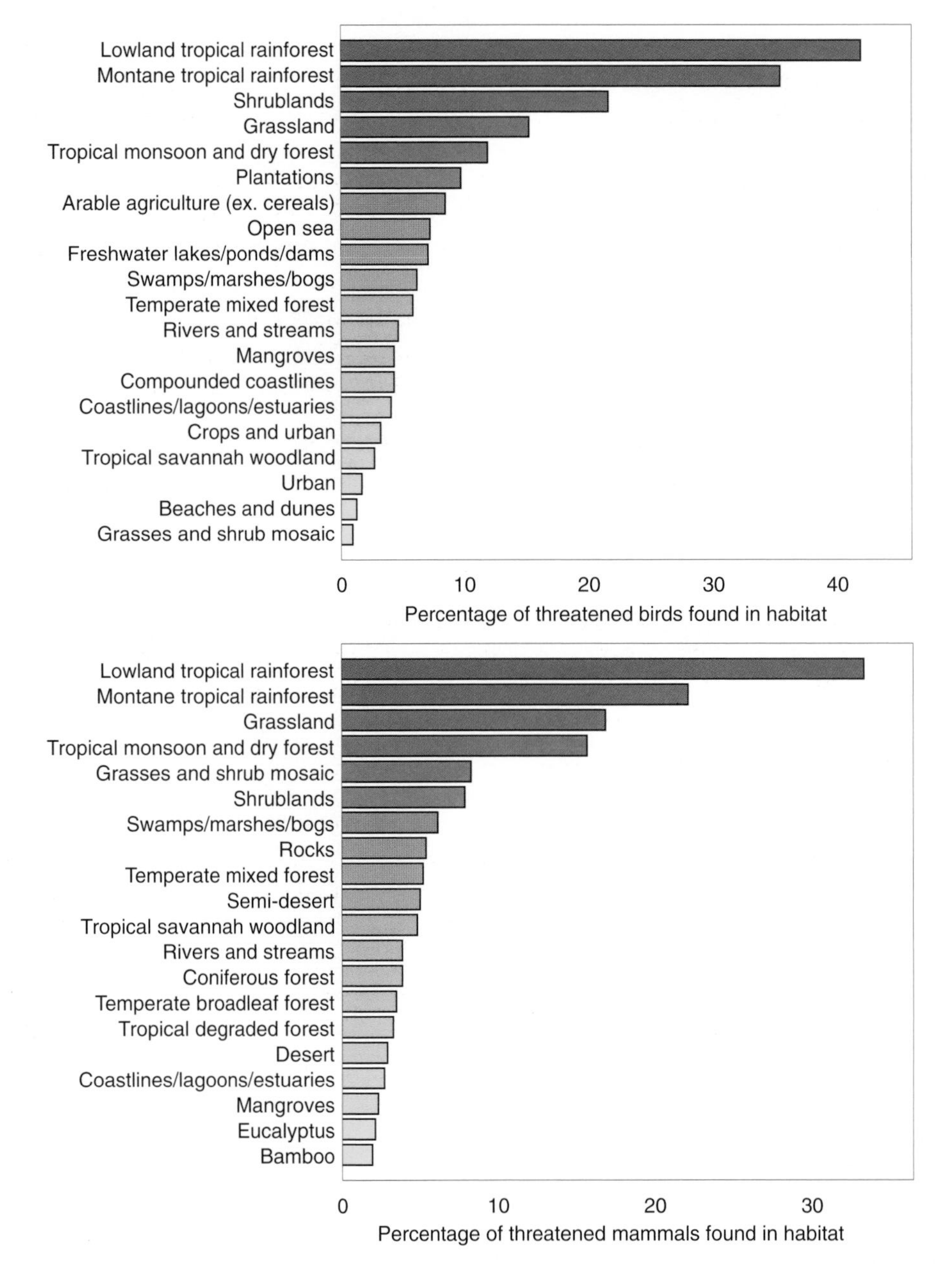

Figure 8. The top twenty habitat types (see Annex 1) with the largest numbers of (above) threatened birds, and (below) threatened mammals (1,180 threatened birds were assigned to habitat types and 515 threatened mammals were assigned).

forest. According to BirdLife, only a few threatened birds (16%) use forests and depend on another habitat type. The figures for mammals are equally compelling with 33% occurring in lowland rain forest and 22% in montane rain forest, but it is not clear what proportion of these mammals are totally dependent on the forests for their survival. Under the BirdLife habitat classification system, occurrence within a habitat is also graded as to whether it is a critical habitat for the species or a minor one. In the system used for the *IUCN Red List*, occurrence is recorded simply as presence/absence only.

The next set of habitats which are important to both birds and mammals, includes grasslands, shrublands, and the tropical monsoon and dry forest. Birds in these habitats are apparently less range-restricted than forest birds (BirdLife International 2000), and this might also apply to mammals.

After the top six categories, the habitat preferences for threatened birds and mammals diverge markedly. It appears that birds are more adaptable and are able to survive more readily in highly transformed habitats such as plantations, agricultural lands and urban areas. Mammals on the other hand (at least those sampled) appear to be far less tolerant of such transformed habitats and disturbance, and none of these types of habitat feature in the top 20. Wetlands and other freshwater aquatic habitats are important for a small number of threatened birds and mammals. These freshwater habitats are more important than coastal and estuarine habitats in both birds and mammals.

Open sea is a very important habitat for many pelagic birds (7% of the threatened species). This habitat is also important for marine mammals, but very few of these species were documented for the 2000 Red List. As a result, this category does not appear in the top twenty habitats for mammals due to a sampling bias.

Drier habitats such as semi-deserts, deserts, rocks and even temperate and coniferous forests are all utilized to some degree (in some cases exclusively) by a variety of threatened mammals. Threatened birds generally appear to avoid such areas. There are also some highly specialized habitats like bamboo and eucalyptus forests, which are important for a small number of threatened mammals, but which are not important for threatened birds.

This analysis shows that there are some key habitats common to both threatened birds and threatened mammals, and these constitute clear conservation priorities. Conservation of extensive areas of tropical rainforests is essential if we are to prevent the total loss of a large number of bird and mammal species, most of which are totally dependent on this habitat for survival. Perhaps more surprisingly, it is also important to prioritize grasslands, shrublands and savannas if mammalian and avian species diversity is to be maintained. Broadly-based conservation strategies are needed to focus on these habitats, and these should be designed to benefit many other species, not just the threatened birds and mammals. A major challenge lies with the threatened species that occur in other less frequently used habitats. When looking at marine birds, birds in artificial habitats, and mammals in semi-deserts, deserts or temperate areas, separate conservation strategies, often developed in addition to other conservation activities, may have to be investigated.

An overview of the major threats

The SSC Species Information Service, through consultation with the SSC membership and SSC's partners, has compiled a hierarchical list of categories describing the many threats to a species (see Annex 5). These threat types are being tested for the first time through their application in documenting the major threats to species on the Red List. The system was applied to 720 mammal species (64% of those threatened), to 1,173 threatened bird species (almost all of those threatened) and to 2,274 (41%) of the threatened plants (primarily trees). For the purposes of this analysis the lower level threats in the hierarchy (see Annex 5) were grouped together to identify the major threats at the primary level and then two of the most important threats were further analysed to identify the specific threatening processes involved. As with the habitats analysis, a species can have more than one threat and so the values within a species group are non-exclusive.

Habitat loss and degradation

The results in Figure 9 show that the most pervasive and over-riding threat to birds, mammals and plants is habitat loss and degradation, affecting 89% of all threatened birds, 83% of the threatened mammals sampled and 91% of the threatened plants sampled. The three primary causes of habitat loss (Figure 10) are agricultural activities (a category which includes crop and livestock farming, and timber plantations), extraction activities (which includes mining, fisheries, logging, and harvesting), and development (which includes human settlements, industry and all the associated infrastructure like roads, dams, power lines, etc.). Agricultural activities affect 827 threatened bird species (70% of all), 1,121 plant species (49% of all) but surprisingly, only 92 (13% of all) of the threatened mammals. Extraction activities had the most impact on plants with 1,365 threatened species being affected (60% of all); 622 threatened birds (53% of all) were also affected. According to BirdLife International (2000), selective logging alone impacts 31% of threatened bird species. Extraction appears to have minimal impact on mammals. Developmental activities affect 769 of the threatened plants (34% of all), 373 birds (32% of all) and only 59 threatened mammals (8% of all). The fourth cause of habitat loss is given as 'unspecified causes', and 495 (69% of all) mammals were classified under this. But using the information from the levels lower down in the hierarchical system, two key threats to mammals are fragmentation (6% of all species) and deforestation (9% of all species) both of which are due to unspecified causes. Clearly, when the threats to mammals are better documented, it is likely that the causes of habitat loss will be more comprehensively attributed to agricultural activities, extraction and development, thus possibly leading to results that are more similar to those for birds and plants. In an analysis of the threats faced by threatened species in the USA (Wilcove *et al.* 2000), habitat loss and degradation emerged as the greatest threat affecting more than 80% of threatened species.

Direct loss and exploitation

Figure 9 shows that direct loss and exploitation also has a major impact on birds (37% of all), mammals (34% of all) and plants (8% of all). This category can be broken down into hunting and collecting activities and the impacts of trade (both legal and illegal) (Figure 10). This shows that 338 threatened bird species (28% of all) are impacted by hunting and collecting, 212 mammals (29% of all) and 169 plants (7% of all). Trade impacts 13% of both threatened birds and threatened mammals, while less than 1% of the threatened plants were impacted by trade.

Alien invasives

Under the hierarchical scheme in Annex 5, alien invasives are placed at level two, but from the results obtained, they are a significant direct threat and should rather be considered at level one. Invasive species are also an important threat affecting 350 (30%) of all threatened birds, and 361 threatened plant species (15% of all). This threat appears to have less impact on mammals, affecting only 69 species (10% of all). The extinction of most bird species since 1800, especially those on islands, is largely attributed to the introduction of alien invasive species (BirdLife International 2000). Therefore it is alarming to note that 30% of threatened birds are currently being affected by invasive species.

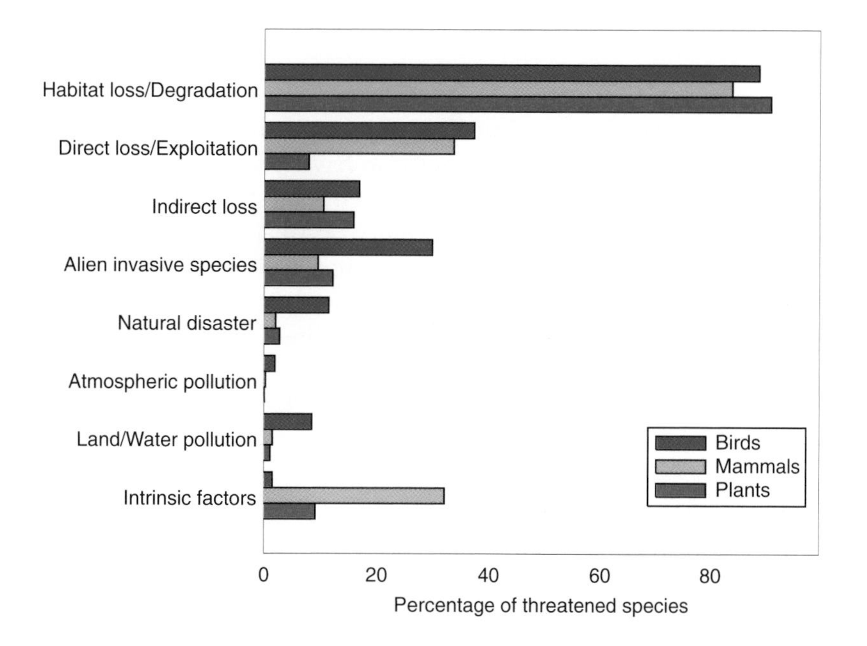

Figure 9. Major threats to threatened birds, mammals and plants, using level one categories of threat from Annex 5 and moving alien invasive species to level one. Habitat loss is the greatest threat to the 1,173 threatened birds, 720 threatened mammals and 2,274 threatened plants sampled.

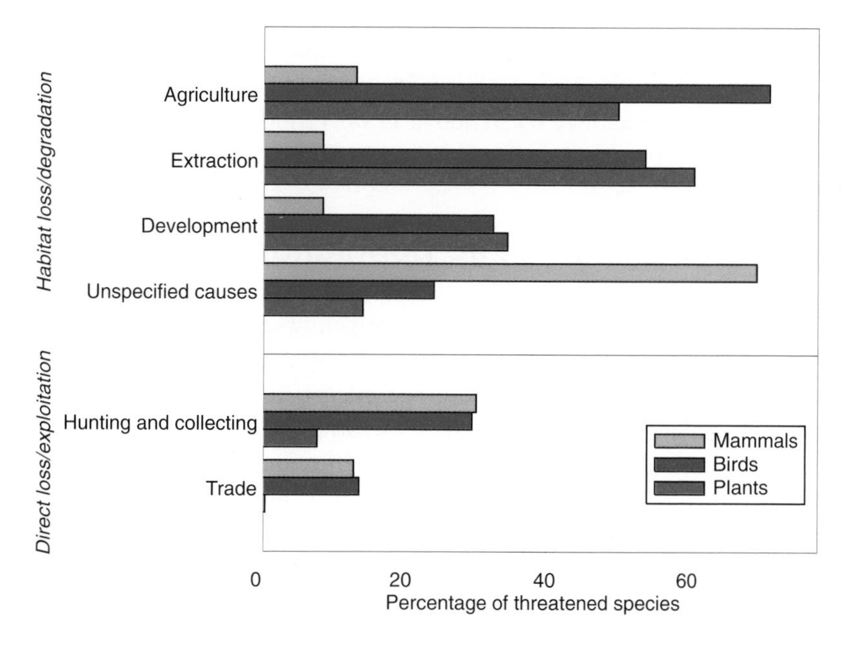

Figure 10. The level one threat habitat loss and degradation is subdivided into various forms as represented by the first four categories, while the last two categories (hunting and collecting and trade) are subdivisions of the direct loss and exploitation threat in level one. The numbers of threatened birds, mammals and plants in each sub-category is a percentage of the total number of species in the group sampled (i.e. 1,173, 720 and 2,274).

Intrinsic factors

Factors such as poor dispersal, poor recruitment, high juvenile mortality and inbreeding have been grouped together as intrinsic factors and they appear to have a relatively important impact on 231 threatened mammal species (32% of all) and 208 threatened plant species (9% of all), whereas only 18 threatened birds were scored under this threat category. Further examination of the original data shows that species with highly restricted distributions were usually placed into this category (i.e. the species were inherently very rare). It is possible that the number of bird species at risk from intrinsic factors has been under-recorded, and this might be clarified as documentation becomes more comprehensive.

Other threats

Of the remaining threats in Figure 9, natural disasters have some impact as do land and water pollution. But as these are largely episodic events rather than ongoing factors, they are unlikely to affect large numbers of threatened species, as their impact will in most cases be limited both in space and time. Very few species were recorded as being threatened by atmospheric pollution, a category that includes factors like global warming, acid rain precipitation and ozone hole effects. However, the impacts of these more insidious threats may be difficult to detect. In addition they are most likely to be more prevalent among other groups of taxa, especially amphibians and reptiles, which appear to be more sensitive to such global changes. It should be noted that the results presented here are only for threatened birds and a selection of threatened mammals and plants, and when these analyses are performed for a broader variety of species, some changes might be expected. This is partly confirmed by the analysis of threats in the USA by Wilcove *et al.* (2000), which found that for threatened freshwater organisms, pollution was the second most important threat after habitat loss.

Threatened subspecies and sub-populations

Countless species, although not yet globally threatened, now exist in reduced numbers as highly fragmented and isolated sub-populations, and many of these sub-populations are threatened with extinction. Frequently these sub-populations are accorded taxonomic recognition in that they are named as subspecies or, in the case of many plants, as varieties. The IUCN Red List Categories and Criteria may be equally applied to such subspecies, varieties and even unnamed but geographically isolated sub-populations (also referred to as 'stocks'). The increasing concern about the loss of genetic diversity through the local extinction of infra-specific taxa and sub-populations is focusing more attention on the conservation assessment and listing of these taxa.

Although the name *IUCN Red List of Threatened Species* implies that the primary focus is at the species level, the 2000 Red List also includes all assessments that have been made at the infra-specific and sub-population levels. The analysis of the 2000 Red List has focussed only at the species level, but Table 7 has been included as any indication of the growing number of infra-specific and sub-population listings. The 2000 Red List includes 1,769 infra-specific taxa and sub-populations, of which 1,237 are listed in the threatened categories (524 vertebrates and 638 plants) and 52 are Extinct or Extinct in the Wild.

Table 7. Summary of infra-specific taxa and sub-populations on the 2000 Red List

Class	EX	EW	Subtotal	CR	EN	VU	Subtotal	LR/cd	LR/nt	DD	Total
Vertebrates											
Mammals	27	3	**30**	96	150	176	**422**	42	110	145	**749**
Birds	0	0	**0**	0	0	0	**0**	0	0	0	**0**
Reptiles	1	2	**3**	15	14	14	**43**	0	1	1	**48**
Amphibians	0	0	**0**	1	1	3	**5**	0	0	2	**7**
Fish	2	0	**2**	12	22	20	**54**	3	6	2	**67**
Subtotal	**30**	**5**	**35**	**124**	**187**	**213**	**524**	**45**	**117**	**150**	**871**
Invertebrates											
Insects	0	0	**0**	4	18	12	**34**	0	1	1	**36**
Molluscs	6	5	**11**	5	5	11	**21**	0	8	35	**75**
Crustaceans	0	0	**0**	1	5	14	**20**	0	0	0	**20**
Others	0	0	**0**	0	0	0	**0**	0	0	0	**0**
Subtotal	**6**	**5**	**11**	**10**	**28**	**37**	**75**	**0**	**9**	**36**	**131**
Plants											
Mosses	0	0	**0**	0	0	0	**0**	0	0	0	**0**
Gymnosperms	0	0	**0**	3	11	47	**61**	4	15	4	**84**
Dicotyledons	6	0	**6**	75	109	384	**568**	10	66	19	**669**
Monocotyledons	0	0	**0**	3	3	3	**9**	2	2	1	**14**
Subtotal	**6**	**0**	**6**	**81**	**123**	**434**	**638**	**16**	**83**	**24**	**767**

The reassessment of the primates for the 2000 Red List paid a great deal of attention to subspecies, as there was a strong feeling that conservation actions on the ground could often be targeted more effectively, if it was known for example, that a particular subspecies of primate occurring in a single isolated forest patch was listed in one of the threatened categories. This was particularly important in those instances where the species as a whole was considered less threatened and hence conservation actions might not be taken for any highly threatened subspecies unless attention was specifically drawn to them. Of the 749 mammal taxa indicated in Table 7, 239 are primate subspecies, 148 of which are listed as globally threatened. This example illustrates the extent of the potential problem of ignoring such taxa, and it is hoped that the inclusion of these taxa on the Red List will lead to better-informed conservation planning. The policy of the SSC Red List Programme, however, is that the *IUCN Red List* should remain focussed primarily at the species level.

Recording extinctions

The SSC Red List Programme Office frequently receives requests to provide information on how many species have gone extinct in the last 100 years. It is extremely difficult to answer such requests because of the problems in recording contemporary extinction events. It is frequently stated that species are being lost every day, particularly invertebrates and other small cryptic organisms, which have not as yet been discovered or named. Even if they have been discovered and named, they are often too small to be noticed without special sampling procedures, or they occur in remote areas where regular monitoring is impossible. The process of decline and eventual extinction may take place over many years or even centuries in the case of very long-lived organisms like some of the large mammal and tree species. The terminal stages in the process of extinction are in fact seldom observed, except in cases where extreme events (e.g., the excessive hunting of the Passenger Pigeon *Ectopistes migratorius*, or the mass extinction of native snails in French Polynesia and Hawaii following the introduction of the predatory snail *Euglandina rosea* to Pacific islands) have very noticeable impacts, and can then, as in the latter case, be the subject of intensive study. For example, the Pacific Island Land Snail Group was established partly under the auspices of SSC's Conservation Breeding Specialist Group (CBSG) to monitor the impacts of the introduced snail and to establish *ex situ* breeding populations for the remaining native species.

In most cases, it takes many years before the lack of sightings of a species generate sufficient concern to stimulate active searches, and even then, it could take many more years before sufficient negative evidence has accumulated to be able to pronounce that a species is Extinct. The criterion for extinction in the IUCN Red List Categories is that "A taxon is presumed extinct in the wild when exhaustive surveys in known and/or expected habitat, at appropriate times (diurnal, seasonal, annual) throughout its historic range have failed to record an individual. Surveys should be over a time frame appropriate to the taxon's life cycle and life form" (IUCN 1994, p. 14). For an ephemeral orchid species, that only appears in the first year after a fire and which grows in a habitat where fires only occur approximately once every 25 years, it could take a very long time to gather sufficient evidence to support an Extinct listing.

From the problems outlined above, it is clear that it is virtually impossible to state with any precision how many species have gone extinct, never mind give a precise date of when the extinction occurred. Likewise it is very difficult to predict with any certainty how many species, let alone which ones, will become extinct in the next one hundred years. These problems are further illustrated by the occasional rediscovery of species thought to be Extinct. The Vietnamese Warty Pig *Sus bucculentus*, was described from two skulls in the late 1800s, as the species was not subsequently recorded in any surveys it was treated as possibly extinct in the 1994 IUCN Red List and finally as Extinct in 1996. In 1997, a report was published indicating that the species still existed based on the identification of a partial fresh skull, which was obtained from some hunters (Groves *et al.* 1997). The area where it was collected is the Annamite Range in Lao PDR, an area that is now famous because of the recent discovery of new and previously undescribed large mammals. Another example, is a species of rice rat *Nesoryzomys fernandinae*, found on the Galápagos island of Fernandina. Although only described in 1980, this species was listed as Extinct in 1996 because no living specimens had been collected. Thorough field surveys of the islands have since shown that *N. fernandinae* still exists, although it is threatened as a result of the introduction of the alien black rat *Rattus rattus*, a species responsible for the extinction of a number of other native rodent species in the Galápagos (Dowler *et al.* 2000).

In response to the problems of compiling an authoritative list of recent extinctions a Committee on Recently Extinct Organisms (CREO) was established by the Center for Biodiversity and Conservation at the American Museum of Natural History (http://creo.emnh.org/index.html). CREO has developed a set of criteria which are used to assess allegedly extinct species and assign them to the list of resolved extinctions or to one of several lists of unresolved extinctions, which are based on increasing degrees of uncertainty. In order to assign species to these lists, CREO draws on scientific expertise world-wide, including the help and knowledge of many SSC members. The criteria for the list of resolved extinctions are very similar to those for the IUCN Extinct Category, hence the Red List Programme has agreed to collaborate with CREO in the compilation of this list.

The list of extinctions in the 2000 Red List incorporates the updated and well-documented list of bird extinctions since 1500 AD, compiled by Thomas Brooks for CREO and which has also been incorporated into *Threatened Birds of the World* (BirdLife International 2000). The previous list of bird extinctions was largely unsubstantiated and outdated and included a number of species which certainly went extinct before 1500 AD (e.g., the giant moas) and are often only known from sub-fossil remains. The revised list also excluded several undescribed species from the Mascarene Islands and a number of others, which are now considered to have been only subspecies of still extant species. Despite these removals, the number of Extinct birds has increased substantially from 104 in 1996 to 128 in 2000, plus there are an additional four species which are Extinct in the Wild (Table 8). This increase is due to better documentation and new knowledge, but we now know from this that 103 of these extinctions have occurred since 1800, indicating an extinction rate 50 times that of the background rate (BirdLife International 2000).

For the other vertebrates, apart from the mammals, there has not been much change (Table 8). The changes in the mammals are partly a result of the attempt to resolve differences between the provisional CREO list and the IUCN list (MacPhee and Flemming 1999) and further changes as a result of these discussions are likely to occur in the future. An area which has not yet been examined is the enormous discrepancy between the provisional list of fish extinctions compiled by CREO (Harrison and Stiassny 1999), which indicates only three resolved extinctions whereas the 2000 Red List includes 81 Extinct fish species (Table 8).

Among the invertebrates, the only changes were in the molluscs (Table 8). Of the 303 recorded molluscan extinctions, only four are of marine molluscs. The land and freshwater molluscs therefore emerge as the faunal compartment most prone to extinction. The total of 303 species (0.04% of all described molluscs) represents only a fraction of mollusc extinctions world-wide, since many poorly known taxa or regions have not been evaluated. For instance, the "apparent" increase in extinctions since 1996 is largely related to proper surveys and documentation of species in the USA. In the 1996 Red List, these were listed as Critically Endangered, whereas in 1994 they had been listed as possibly extinct. In contrast, there are a few rare cases of species believed to be extinct being rediscovered in the last four years. The Madeiran endemic land snail, *Discus guerinianus*, has been rediscovered at the far, western end of the island. This species had not been recorded since the 1870s at the original locality near Funchal, despite intensive survey efforts between 1983 and 1996. The new discoveries at two remote, closely adjacent sites show that small pockets of habitat are sufficient to maintain land-snail species under certain conditions (Cameron and Cook, 1999).

An analysis of the plant extinctions is premature, as the 73 species listed only include trees. All the plants listed as extinct or possibly extinct in the *1997 IUCN Red List of Threatened Plants* need to be re-assessed using the extinction criteria, before any analysis can be done.

The 816 recorded extinctions since 1500 AD are not evenly distributed around the world as can be clearly seen in Figure 11 which shows the top twenty countries with the highest number of extinctions. Fifteen of these countries or territories are islands and even in the USA, the country with by far the greatest number of recorded extinctions, the vast majority are from the Hawaiian islands. The greater vulnerability of island species to extinction is readily understood because of their limited ranges, usually smaller population sizes and because in most cases they evolved in the absence of certain pressures like predators and competitors. The introduction of aggressive alien species and various predators since the start of European exploration and settlement of most islands in the 1500s has had a devastating effect on the native biota of many islands. The emphasis on islands may be slightly biased

Table 8. Recorded number of extinctions by species group in 1996 and 2000

	1996			2000		
	EX	EW	Total	EX	EW	Total
Vertebrates						
Mammals	86	3	89	83	4	87
Birds	104	4	108	128	3	131
Reptiles	20	1	21	21	1	22
Amphibians	5	0	5	5	0	5
Fishes	81	11	92	81	11	92
Subtotal	**296**	**19**	**315**	**318**	**19**	**337**
Invertebrates						
Insects	72	1	73	72	1	73
Molluscs	230	9	239	291	12	303
Crustaceans	9	1	10	8	1	9
Others	4	0	4	4	0	4
Subtotal	**315**	**11**	**326**	**375**	**14**	**389**
Plants						
Mosses				3	0	3
Gymnosperms				0	1	1
Dicotyledons				69	14	83
Monocotyledons				1	2	3
Subtotal				**73**	**17**	**90**
Total				**766**	**50**	**816**

Note: EX is Extinct and EW is Extinct in the Wild. The counts are of species known to have become globally Extinct since 1500 AD, although some of the mammals included are under dispute as they are allegedly only known from fossil evidence. There is no strictly comparable baseline for the plants as the extinction criteria were different for the 1997 plants Red List and the figures now include non-trees so they cannot be compared to *The World List of Threatened Trees*. The extinction data is in the process of being updated, so that there is some concordance between the IUCN list of extinctions and the list of resolved extinctions being compiled by the Committee for Recently Extinct Organisms (CREO).

because islands are relatively smaller and easier to survey than large continental areas, making it somewhat easier to conclude that an island species has become extinct.

Conclusions

The *2000 IUCN Red List of Threatened Species*, like all its predecessors, makes for extremely depressing reading. The Red List documents the loss of 816 species in the last 500 years due to the impact of human activities, no doubt with many more that are being lost which are not even known about. A further 11,046 threatened species are considered to be facing a high risk of extinction in at

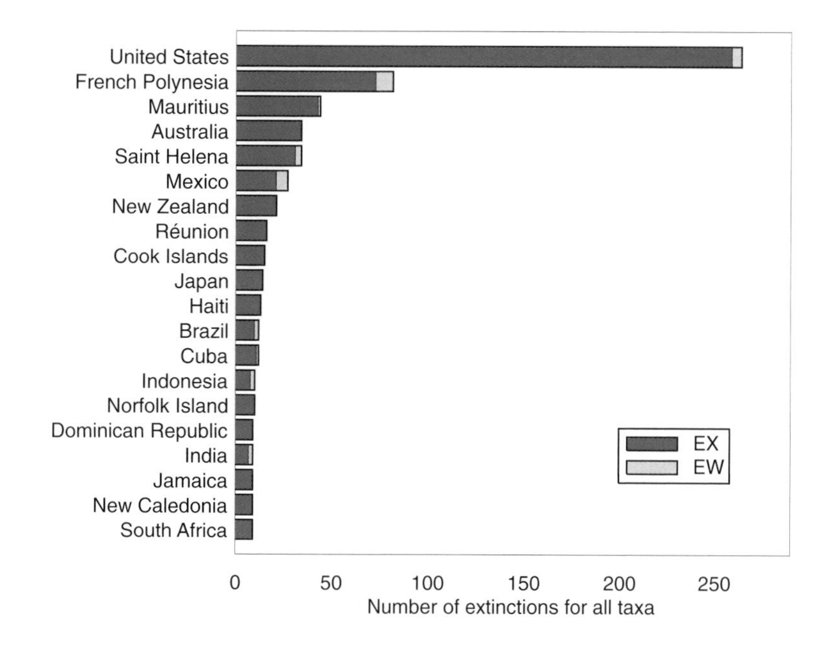

Figure 11. The twenty countries with the highest number of extinctions since 1500 AD. EX = Extinct and EW = Extinct in the Wild.

least the near future, as a result both directly and indirectly of human activities. This includes 24% of all mammal species, 12% of all bird species and 16% of all conifer species. A further 4,595 species are on the brink of moving into one of the threatened categories unless something is done soon to reverse their continued population declines which are once again the result of human activities. Apart from the sheer numbers of species listed as being threatened or potentially threatened, the rapid movement of many species of mammals and birds through the threatened categories towards being Critically Endangered in just the last four years is extremely alarming. Humans are the cause of this deteriorating situation, which is increasingly being referred to as the sixth extinction crisis. But people also have the responsibility and the ability to reverse the situation. To do this will take both knowledge and commitment.

The 2000 Red List provides some of the basic knowledge which can be combined with other information so that it can be used by conservation planners and decision-makers to establish priorities and take appropriate remedial actions. As the database of information upon which the Red List is based improves, so our capability to produce more sophisticated analyses of priorities and different scenarios will increase. However, all this knowledge needs to be transformed into action. And effective action requires much deeper commitment. At the beginning of this book, Russell Mittermeier challenges the global community to "increase levels of support and commitment" to conservation. He goes on to say "We need to act decisively, we need to act now, and we need to act at a scale far beyond anything that has ever been done before. The findings in the 2000 Red List, together with conservation priority setting exercises . . . provide us with a solid foundation as to where to focus. We need to mobilize the human and financial resources at a level at least one and more likely two orders of magnitude beyond anything previously realized."

References

Akçakaya, H.R. and Ferson, S. 1999. *RAMAS Red List: Threatened Species Classifications Under Uncertainty*. Version 1.0. Applied Biomathematics, New York.

Baillie, J. and Groombridge, B. (compilers and editors) 1996. *1996 IUCN Red List of Threatened Animals*. IUCN, Gland, Switzerland and Cambridge, UK.

Bennett, P.M. and Owens, I.P.F. 1997. Variation in extinction risk among birds: chance or evolutionary predisposition? *Proceedings of the Royal Society of London* B 264: 401–408.

BirdLife International 2000. *Threatened Birds of the World*. Lynx Edicions and BirdLife International, Barcelona and Cambridge, UK.

Brummitt, R.K. 1992. *Vascular Plant Families and Genera*. Royal Botanic Gardens, Kew.

Brummitt, R.K. and Powell, C.E. 1992. *Authors of Plant Names*. Royal Botanic Gardens, Kew.

Cable, S. and Cheek, M. 1998. *The Plants of Mount Cameroon. A Conservation Checklist*. Royal Botanic Gardens, Kew.

Cameron, R.D.C. and Cook, L.M. 1999. Island land snail relocated. *Journal of Molluscan Studies* 65: 273–274.

Collar, N.J., Crosby, M.J. and Stattersfield, A.J. 1994. *Birds to Watch 2: the World List of Threatened Birds*. BirdLife Conservation series No. 4. BirdLife International, Cambridge, UK.

Cronquist A. 1981. *An Integrated System of Classification of Flowering Plants*. Columbia University Press, New York.

Cronquist, A. 1988. *The Evolution and Classification of Flowering Plants*. 2nd edition. New York Botanical Garden, Bronx, New York.

Daugherty, C.H., Cree, A., Hay, J.M. and Thompson, M.B. 1990. Neglected taxonomy and continuing extinctions of tuatara (*Sphenodon*). *Nature* 347: 177–179.

Dowler, R.C., Carroll, D.S. and Edwards, C.W. 2000. Rediscovery of rodents (Genus *Nesoryzomys*) considered extinct in the Galápagos Islands. *Oryx* 34: 109–117.

Duellman, W.E. 1993. *Amphibian Species of the World: additions and corrections*. University of Kansas Museum of Natural History, Special Publication No. 21.

Eschmeyer, W.N. 1990. *Catalog of the Genera of Recent Fishes*. California Academy of Sciences, San Francisco.

Eschmeyer W.N. 1998. *Catalog of Fishes*. California Academy of Sciences, San Francisco.

Farjon, A. 1998. *World Checklist and Bibliography of Conifers*. World Checklists and Bibliographies, 3. Royal Botanic Gardens, Kew.

Farjon, A. and Page, C.N. 1999. *Conifers. Status Survey and Conservation Action Plan*. IUCN/SSC Conifer Specialist Group. IUCN, Gland Switzerland and Cambridge, UK.

Frost, D.R. 1985. *Amphibian Species of the World: Taxonomic and Geographical Reference*. Allen Press Inc. and the Association of Systematics Collections, Lawrence, Kansas. I-V, 1–732.

Glaw, F. (Zoologische Staatssammlung München, Munchhausenstr. 21, D–81247, München, Germany) and Kohler, J. (Zoologisches Forschungsinstitut und Museum Alexander Koenig, Adenauerallee 160, D–53113

Bonn, Germany). 1998. *Amphibians of the World*. Update of Duellman, W. E. 1993. *Amphibian Species of the World: additions and corrections*.

Groves, C.P., Schaller, G.B., Amato, G. and Khounboline, K. 1997. Rediscovery of the wild pig *Sus bucculentus*. *Nature* 386: 335.

Hallingbäck, T. and Hodgetts, N. (compilers) 2000 in press. *Mosses, Liverworts and Hornworts. Status Survey and Conservation Action Plan for Bryophytes.* IUCN/SSC Bryophyte Specialist Group. IUCN, Gland, Switzerland and Cambridge, UK.

Hammond, P.M. 1995. The current magnitude of biodiversity. In: V.H. Heywood (ed.) *Global Biodiversity Assessment*, pp. 113–138. Cambridge University Press, Cambridge.

Harrison, I.J. and Stiassny, M.L.J. 1999. The quiet crisis. A preliminary listing of the freshwater fishes of the world that are extinct or "missing in action". In: R.D.E. MacPhee (ed.) *Extinctions in Near Time*, pp. 271–331. Kluwer Academic/Plenum Publishers, New York.

Hollis, S. and Brummitt, R.K. 1992. *World Geographic Scheme for Recording Plant Distributions*. Plant Taxonomic Database Standards No. 2, International Working Group on Taxonomic Databases for Plant Sciences (TDWG). Hunt Institute for Botanical Documentation, Pittsburgh.

Hudson, E. and Mace, G. (eds) 1996. *Marine Fish and the IUCN Red List of Threatened Animals*. Report of a workshop held in collaboration with WWF and IUCN at the Zoological Society of London, April 29th-May 1st 1996.

ISO 1997. ISO–3166–1. *Codes for the representation of names of countries and their subdivisions – Part 1: Country Codes*. Fifth edition. ISO 3166 Maintenance Agency at DIN, Berlin.

IUCN 1994. *IUCN Red List Categories*. Prepared by the IUCN Species Survival Commission. IUCN, Gland, Switzerland.

Iverson, J.B. 1992. *A Checklist with Distribution Maps of the Turtles of the World*. Second edition. Published by the author, Richmond, Indiana.

King, F.W. and Burke, R.L. 1989. *Crocodilian, Tuatara and Turtles Species of the World: a taxonomic and geographical reference*. The Association of Systematics Collections, Lawrence, Kansas. I-V.

Mabberley, D.J. 1997. *The Plant-Book. A portable dictionary of the higher plants.* Second edition. Cambridge University Press, Cambridge.

Mace, G.M. and Balmford, A. 2000. Patterns and processes in contemporary mammalian extinction. In: A. Entwistle and N. Dunstone (eds) *Priorities for the Conservation of Mammalian Diversity. Has the Panda had its day?*, pp. 28–52. Cambridge University Press, Cambridge.

Margulis, L. and Schwartz, K.V. 1988. *Five Kingdoms: an Illustrated Guide to the Phyla of Life on Earth*. 2nd edition. W.H. Freeman and Company, New York.

Master, L.L., Stein, B.A., Kutner, L.S. and Hammerson, G.A. 2000. Vanishing assets. Conservation status of US species. In: B.A. Stein, L.S. Kutner, and J.S. Adams (eds) *Precious Heritage. The status of biodiversity in the United States*, pp. 93–118. Oxford University Press, Oxford, New York.

May, R.M., Lawton, J.H. and Stork, N.E. 1995. Assessing extinction rates. In: J.H. Lawton and R.M. May (eds) *Extinction Rates*, pp. 1–24. Oxford University Press, Oxford.

MacPhee, R.D.E. and Flemming, C. 1999. *Requiem Æternum*. The last five hundred years of mammalian species extinctions. In: R.D.E. MacPhee (ed.) *Extinctions in Near Time*, pp. 333–371. Kluwer Academic/Plenum Publishers, New York.

Morony, J.J., Bock, W.J. and Farrand, J.J. 1975. *Reference List of the Birds of the World.* Department of Ornithology, American Museum of Natural History, New York.

Nowak, R.M. 1999. *Walkers Mammals of the World.* Sixth edition. John Hopkins University Press, Baltimore.

Oldfield, S., Lusty, C. and MacKinven, A. 1998. *The World List of Threatened Trees*. World Conservation Press, Cambridge.

Parker, S.B. 1982. *Synopsis and Classification of Living Organisms*. McGraw-Hill, New York.

Purvis, A., Agapow, P.-M., Gittleman, J.L. and Mace, G.M. 2000. Nonrandom extinction and the loss of evolutionary history. *Science* 288: 328–330.

Rylands, A.B., da Fonseca G.A.B., Leite, Y.L.R. and Mittermeier, R.A. 1996. Primates of the Atlantic forest. Origin, distributions, endemism, and communities. In: M.A. Norconk, A.L. Rosenberger and Garber, P.A. (eds.) *Adaptive Radiations of Neotropical Primates*, pp. 21–51. Plenum Press, New York.

Sibley, C.G. and Monroe, B.L. Jr. 1990. *Distribution and Taxonomy of Birds of the World.* Yale University Press, New Haven.

Sibley, C.G. and Monroe, B.L. Jr. 1993. *A Supplement to Distribution and Taxonomy of Birds of the World.* Yale University Press, New Haven.

Stein, B.A., Adams, J.S., Master, L.L., Morse, L.E. and Hammerson, G.A. 2000. A remarkable array. Species diversity in the United States. In: B.A. Stein, L.S. Kutner, and J.S. Adams (eds) *Precious Heritage. The status of biodiversity in the United States*, pp. 55–92. Oxford University Press, Oxford, New York.

Uetz, P. and Etzold, T. 1996. The EMBL/EBI Reptile Database. *Herpetological Review* 27: 174–175.

Walter, K.S. and Gillett, H. (eds) 1998. *1997 IUCN Red List of Threatened Plants.* Compiled by the World Conservation Monitoring Centre. IUCN, Gland, Switzerland and Cambridge, UK.

Wilcove, D.S., Rothstein, D., Dubow, J., Phillips, A. and Losos, E. 2000. Leading threats to biodiversity. What's imperilling U.S. species. In: B.A. Stein, L.S. Kutner, and J.S. Adams (eds) *Precious Heritage. The status of biodiversity in the United States*, pp. 239–254. Oxford University Press, Oxford, New York.

Wilson, D.E. and Reeder, D.M. (eds) 1993. *Mammal Species of the World a Taxonomic and Geographic Reference.* Second edition. Smithsonian Institution Press, Washington and London.

World Conservation Monitoring Centre 2000. *Global Biodiversity: Earth's living resources in the 21st century.* By: Groombridge, B. and Jenkins, M.D. World Conservation Press, Cambridge.

WWF and IUCN 1994. *Centres of Plant Diversity. A guide and strategy for their conservation.* 3 volumes. IUCN Publications Unit, Cambridge, UK.

Annex 1. Recent Developments in the IUCN/SSC Red List Programme

New developments

In order to achieve the goals and objectives of the Red List Programme as specified in the Introduction, a number of new developments have become necessary. These changes will help to maintain the high profile and scientific integrity of the *IUCN Red List*.

Establishment of Red List Authorities

The improved objectivity of the 1994 IUCN Red List Categories and Criteria revealed that the previously *ad hoc* process of including species on the Red List had to be changed. To achieve this, Red List Authorities are being established for all taxonomic groups included on the *IUCN Red List*. In most cases, the Red List Authority (RLA) is the SSC Specialist Group responsible for the species, group of species or specific geographic area. An exception is birds, where BirdLife International is the designated RLA. In cases where the SSC and its partner networks do not cover a particular taxonomic group or geographic region, the Red List Programme will recommend the appointment of other appropriate organizations or networks as Red List Authorities or National Red List Advisory Groups. The latter will also form a much-needed link between the many national Red List initiatives and the IUCN Red List.

The role of the Red List Authorities is to ensure that all species within their jurisdiction are correctly evaluated against the IUCN Red List Categories at least once every ten years and, if possible, every five years. These evaluations should include all the necessary documentation as required by set terms of reference and should be done in as consultative manner as is possible. The intention is that no new species will be included on the IUCN Red List until it has been evaluated by two members of an appointed Red List Authority or by the Red List Standards Working Group. In cases where Red List Authorities have over-lapping jurisdictions, no RLA has precedence over another and both authorities are required to collaborate in evaluating the status of the species concerned.

This system places greater responsibility on the SSC network and its partners to ensure that what appears on the IUCN Red List is credible and scientifically accurate.

RAMAS® Red List Software

To help the Red List Authorities conduct all the necessary evaluations, the Red List Programme has adopted the use of RAMAS® Red List, a software package developed by Applied Biomathematics, an ecological software development group based in New York. This software applies the rules of the IUCN Red List Criteria to obtain an assessment, and also includes an algorithm for explicitly handling any data uncertainty. The Red List Programme Subcommittee approved the use of the software on a trial basis for the 2000 Red List. Several new assessments included in the 2000 Red List were made using the software.

Documentation requirements and taxonomic standards

An important shortcoming of the 1996 and 1997 IUCN Red Lists is that the listings were poorly documented and as a result, largely unsubstantiated. The specification of the criteria met provides some justification. However, to rectify this weakness, a new system of minimum documentation requirements is being developed. Many of the new additions to the 2000 Red List or changes in status have been accompanied by some degree of documentation. All species added from now onwards, and any listings that are changed must be documented following the requirements adopted. Red List Authorities are also being encouraged to start documenting all of their taxa on the IUCN Red List. The aim is to get all species on the Red List documented to some degree by the year 2003. The inclusion of these documentation requirements to some extent represent a return to the former Red Data Books produced before the current Red List series began. However, with the move of the Red List to a purely electronic medium, the maintenance and continual updating of such documentation is made much easier. Despite this increased level of documentation, the term 'Red List' will be maintained to avoid any confusion.

Another weakness of the IUCN Red Lists is the lack of sufficiently clear taxonomic standards. Taxonomic standards are being adopted and all species on the Red List should conform to these by the year 2003. All new species' listings, and any revisions to listings, must be in accordance with these taxonomic standards, but deviations are permitted provided they are fully documented and substantiated.

The documentation requirements and taxonomic standards will be reviewed at regular intervals. The documentation and taxonomic validation will make the listing process far more transparent and open to challenge, and will increase its scientific integrity and authority. The entire process will be underpinned by the IUCN/SSC Species Information Service (SIS) to ensure efficient management and integration of relevant data.

Petitions process

With the increasing transparency of the Red List it was decided that a mechanism should also be introduced whereby appeals against current listings could be submitted. A process for this has been established and was launched earlier this year. The listings for four species of marine turtle have been appealed through this mechanism and these appeals will be resolved in time for the 2001 edition of the *IUCN Red List*. Two unofficial petitions were also received about the listing of two tree species. These petitions will be referred to the appropriate Red List Authority in due course.

Criteria Review process

Since 1997, the SSC Red List Programme has been actively engaged in conducting a review of the 1994 IUCN Red List Categories and Criteria in accordance with Resolution 1.4 adopted at the 1st World Conservation Congress held in Montreal, Canada in 1996. The Criteria Review Working Group established for this task has completed its work and the revised set of categories and criteria, with improved definitions, was adopted by IUCN Council in February 2000. The revised system will come into use in 2001 and a summary of the key changes, are described in Annex 7.

Future development of the IUCN Red List as an indicator of biodiversity trends

The driving focus of the SSC Red List Programme over the next few years will be to develop the *IUCN Red List* in a manner that allows a variety of biodiversity indicators to be developed and measured over time. However, in order for credible indicators to be implemented though the Red List, two important changes need to be implemented: a) the further documentation of the Red List as described above and b) the expansion of the taxonomic groups covered in the Red List.

Expanding the coverage of the Red List in a strategic way

In order to develop the biodiversity indices from the *IUCN Red List*, a major expansion of the taxonomic coverage is a very high priority. Using mammals and birds as indicators, it was possible to use the 1996 Red List to draw some preliminary conclusions on the state of degeneration of global biodiversity. However, it was recognized that mammals and birds, on their own, are not necessarily the best indicators of biodiversity. Species that are characteristic of non-terrestrial ecosystems were poorly covered in this assessment. It is important that the Red List Programme is based on continuous monitoring of a selection of higher taxonomic groups that broadly cover and represent the full range of ecosystems world-wide. Ecosystems that are particularly poorly covered at present include: all freshwater and aquatic habitats; all marine, estuarine and coastal habitats; to some extent grasslands, rangelands and deserts; and dead-wood habitats in forests.

In the light of the gaps identified above, the SSC has started the process of identifying priority higher taxon groups that will be selected to obtain the broader ecosystem and taxonomic coverage that is required. Some ambitious targets have already been set, notably complete coverage of the following:

- Amphibians by 2001 (approximately 5,000 species to be assessed)
- Reptiles by 2002 (approximately 8,000 species)
- Freshwater Fish by 2003 (approximately 10,000 species)
- Sharks, Rays and Chimaeras by 2003 (approximately 1,000 species)
- Freshwater Molluscs by 2003 (approximately 5,000 species)

These are very ambitious targets and their achievement depends on the availability of resources. However, even with these groups added, the taxonomic base for the biodiversity indices will still be inadequate. For this reason, the SSC has set in motion processes to identify priority taxonomic groups of plants, invertebrates and marine organisms.

Annex 2. Organization of information

All previous editions of the *IUCN Red List* were produced in book format. The 2000 Red List therefore marks a radical departure from the traditional approach. The reasons for this departure are simple:

- The decision to incorporate plants and animals into a single Red List, rather than treating them separately, means that the total species coverage has more than doubled in 2000, and will increase even more in years to come.
- The 2000 Red List marks the start of a process to document all species listed and this documentation will greatly increase the size but also the utility of the Red List.
- In this electronic era it is time for the *IUCN Red List* to make better use of this medium as it provides a wider audience with easy access to the information.
- The IUCN Red List will be updated on an annual basis from 2000 onwards, and so the production of a book every year would be too prohibitive both in terms of time and cost.

The *2000 IUCN Red List of Threatened Species* not only has a revised title, but is now available as an electronic version on both the World Wide Web and on CD-ROM. The information and layout presented in both versions are identical and are based on a data download from the threatened species database developed and maintained at present by the SSC Red List Programme Office in Cambridge, UK. The database will in due course be transferred to the SSC Species Information Service (SIS) database, which is currently being developed. The information covers all taxa which have been assigned a Red List Category with the exception of those designated as Not Evaluated (NE) or those in the Lower Risk subcategory of least concern. All the assessments presented, except for the 79 geographically isolated sub-population or stock assessments, are of the taxon (species, subspecies or variety) as a whole (i.e. they indicate the global risk of extinction). No national or regional Red List assessments are included, except for an occasional note about this in one of the documentation fields.

The first step in compiling the 2000 Red List was for all the tree assessments made for and published in *The World List of Threatened Trees* (Oldfield *et al.* 1998) to be merged into the same database as all the assessments which appeared in the *1996 IUCN Red List of Threatened Animals*. SSC Specialist Group members were invited twice (in 1999 and in 2000) to submit any new assessments, revisions and corrections. In addition, reports were sent to most animal specialist groups showing them what information was held on their particular species and they were asked to validate this information. A large number of members replied and provided new submissions or corrections for inclusion in the 2000 Red List (see the acknowledgements). Practically all of these submissions were included, but some were held back pending review by one of the established Red List Authorities, or because no documentation was provided with the assessments, which was one of the stipulated requirements for the 2000 Red List. All the new information was entered into the threatened species database. In the limited time left available, the Red List team tried to gather and enter whatever documentation they could on a wide range of species, to enable the analysis presented here to be done. The final task was to incorporate all the bird assessments and documentation provided by BirdLife International into the database.

Documentation of species

The SSC Red List Programme has developed a set of documentation requirements that are being phased in slowly over three years, by the end of which it is hoped that all species on the Red List will be documented to some extent. For the 2000 Red List, people making submissions were requested to provide (in addition to the usual details about name, status, criteria, distribution, etc.) a rationale to support the listing, a list of the major habitats the species was found in, what the major threats were and an indication as to whether the species' population trend was increasing, decreasing, stable or unknown. If the taxon assessed fell within the jurisdiction of an appointed Red List Authority (see Annex 1 on the IUCN/SSC Red List Programme for details) then it was referred to them for evaluation. However, there were difficulties in implementing the system strictly according to the rules laid down so some flexibility was permitted.

The documentation for each species attempts to cover the following:

- Higher taxonomy details recorded including Kingdom, Phylum, Class, Order and Family.
- Scientific name including authority details where ever possible. (Note: for animals the date of description is usually shown, but for plants this is not the case).
- Common names (English, French and Spanish).
- Red List Category and Criteria (only the criteria for the highest category to which the species can be assigned

are specified, not all the other criteria met as is done by BirdLife International in their publications).

- Date of assessment (used to show if a species was last assessed in 1996, 1998, 1999 or 2000).
- An indication if a petition about the status of the species has been lodged.
- An indication if the caveat devised for certain marine fishes applies (see below).
- Countries of occurrence and sub-country units for large countries and islands far from mainland countries (see below).
- Occurrence in marine regions and inland water bodies or systems (see below).
- A rationale for the listing (including any numerical data used, or inferences made, that relate to the thresholds in the criteria).
- Current population trends (where ↑ = improving, ↓ = deteriorating → = stable and ? = uncertain or don't know).
- Major habitat preferences (based on the classification used by the Global Land Cover Characterization (GLCC) with adaptations for freshwater and marine ecosystems – see Annex 4).
- Major threats (using a standard classification of threats developed for the SIS, see Annex 5).
- General notes about population and range, habitat and ecology, threats and what conservation measures have been taken.
- Information on any changes in the Red List status of the species, and why this status has changed.
- Data sources.
- Consultation process (including the name/s of who had made the original assessment, and if a Red List Authority was involved, the names of the individual evaluators and the RLA involved)

The degree of documentation achieved is extremely variable across the list. Very few species have been fully documented in line with the above requirements, but a few Specialist Groups submitted the full documentation required and more. The complete texts of these species accounts will be made available via the SSC web site as they represent a considerable investment of time and energy by the Specialist Group members concerned and are extremely thorough reviews of the species assessed.

An analysis of the key documentation fields indicates that approximately 20% of mammals have been documented, 84% of the birds (for the 2000 Red List the bird documentation only includes the rationale for the listing and not the full text on population, range, habitat, etc., which is only available at present in printed format as *Threatened Birds of the World* (BirdLife International 2000)), 4% of reptiles are documented, 15% of amphibians, only 1% of fish (mainly the sharks and rays), 2% of invertebrates (mainly molluscs), and 91% of plants (mainly from Oldfield *et al.* 1998).

Extinct and Extinct in the Wild species

For these species, extra documentation was required indicating the effective date of extinction, causes of extinction and the details of surveys which have been conducted to search for the species. The starting date for the inclusion of extinctions was previously set at 1600 AD, but this has been moved back to 1500 AD to be in line with the starting date used by CREO. An attempt has been made to collate whatever information is available on each extinct species and this is presented in the various documentation fields.

Subspecies, varieties and sub-populations

Although the name *IUCN Red List of Threatened Species* implies that the primary focus is at the species level, the *IUCN Red List* also includes assessments that are done at the infra-specific or sub-population levels. Ideally, for such taxa to be included in the Red List the global status of the species itself should be assessed. In most instances this is the case and generally these are assessed as Lower Risk least concern, and as so do not appear in the public version. There are some cases, however, especially amongst the plants where this has not been done and it may well be the case that some of these species warrant inclusion in the Red List.

Taxa removed from the 2000 Red List

The Red List is highly dynamic with species moving on and off for a variety of reasons. All these changes are tracked, so that a complete audit trail is kept for each taxon name that ever appears in the Red List. Taxa removed from the 2000 Red List are not shown in a separate list on the web or CD-ROM versions. Requests for further information about these taxa can be directed to the Red List Programme Office (see inside front cover for address details).

Marine caveat

A small number of marine fishes have the letter C in parentheses after the species name details. This indicates that a caveat formulated at the workshop on categorizing marine fishes (Hudson and Mace 1996) applies in particular to these populations. The text of this caveat is reproduced in the following paragraph.

The criteria (A-D) provide relative assessments of trends in the population status of species across many life forms. However, it is recognized that these criteria do not always lead to equally robust assessments of extinction risk, which depend upon the life history of the species. The quantitative criterion (A1abd) for the threatened categories may not be appropriate for assessing the risk of extinction for some species, particularly those with high reproductive potential, fast growth and broad geographic ranges. Many of these species have high potential for population maintenance under high levels of mortality, and such species might form the basis for fisheries.

Distribution information

Distribution is recorded in terms of country names following the 5th edition of the ISO–3166–1 standard (ISO 1997). Unless geographically very remote from each other, islands and other territories are included with the parent country. In the case of species that only inhabit islands significantly distant from the mainland, the island name is given in parentheses (e.g. Spain (Canary Islands)). The naming of such islands follows an updated version of the *World Geographical Scheme for Recording Plant Distributions* (Hollis and Brummitt 1992) which was adopted by the International Working Group on Taxonomic Databases (TDWG). The updated version used, is in fact the second edition that is currently being prepared. The TDWG geographic system also provides a standard set of Basic Recording Units (BRU) which are sub-country units based on provinces or states. The BRU's are used to subdivide very large countries like Australia, Brazil, China, South Africa, the Russian Federation and the United States of America, etc. into smaller more conveniently sized units for recording distributions. This system has been adopted for the 2000 Red List wherever possible. Unfortunately, sub-country information is still lacking for most of the animal species, whereas most of the plants on the Red List have had their distributions recorded down to the BRU level.

With regard to marine species, country records are generally provided only for strictly coastal or inshore (or riverine) cetaceans, sharks and rays, and for other marine species that return to land to breed or nest. Any species without a country name in the database used to generate the web version and CD-ROM, are not included in any of the analyses or tables presented. For some marine species, especially those that are most strictly marine, their distributions are shown as generalized ranges in terms of the FAO Fishing Areas (http://www.fao.org/fi/sidp/htmls/frstmap.htm), indicated (e.g. Atlantic - eastern central) For many inland water species, usually those restricted to a single water system, inland water ranges are also given, with a clear indication from the name if it is a river or a lake. In a few instances, a second river name (e.g. Colorado River, Concho River) indicates a section of the drainage in which the species occurs. In most cases, the countries in which these freshwater species occur are also recorded, but there are a few instances (e.g., cichlid fish in Lake Victoria), where the precise country distribution is unknown, so the only distribution information is the Lake name. Species such as this also do not appear in any of the country analyses or the country table.

In most cases where populations are known to have been introduced or reintroduced to a country, this is indicated by [int] or [re-int] after the country name in the distribution text. Similarly, where populations are known or suspected to have been extirpated from a country, this is indicated by [RE] or [RE?] for Regionally Extinct.

Geopolitical events during recent years may have led to some inconsistency or errors in the distribution information provided. Within reasonable limits every effort has been made to determine which of the new nations that were part of the former Czechoslovakia, Yugoslavia and USSR support species previously attributed to the larger unit. There may be some species in the database which have still not been fully resolved. Adherence to the ISO system also creates some problems as there is often a time lag between a political change and a new ISO code being allocated for the new country. The current version of the ISO codes for example, maintains Hong Kong and Macau as separate units.

The 2000 Red List contains assessments for 79 stocks or geographically isolated sub-populations. In the 1996 Red List these were indicated by (S) after the species name, but this created confusion in some cases, so a geographic name is now given directly after the name e.g. *Eubalaena glacialis* P.L.S. Müller, 1776 (North Pacific stock).

Annex 3. Information sources and quality

The information presented in the 2000 Red List represents an accumulation of knowledge derived from both the *1996 IUCN Red List of Threatened Animals* and *The World List of Threatened Trees*. Readers are therefore referred to both of these previous publications when checking on information sources and data quality. Many of the assessments done for the 1996 list are in the 2000 Red List unchanged and the original sources who provided either the information or the assessments are still recognized. For every species entry, there is a name of one or more assessor or the name of an SSC Specialist Group. In some cases assessments are the product of group discussion, but often they represent the judgement of individual Specialist Group members. In order to ensure greater accuracy and transparency in the listing process, a peer review system of Red List Evaluators has been initiated. As this system was applied for the first time this year, some flexibility was allowed to give Specialist Groups and Red List Authorities time to adjust to the new requirements. The intention of the system is that the assessments of all species on the Red List should be scrutinized and evaluated by at least two people from a designated Red List Authority. The Red List Authorities will be responsible for ensuring that all species they are responsible for are documented and re-assessed at regular intervals.

BirdLife International is the Red List Authority for birds and as such they provided all the bird assessments used for the 2000 Red List. These assessments and the accompanying documentation reflect partially the contents of *Threatened Birds of the World* (BirdLife International 2000).

All mammal species (as listed in Wilson and Reeder, 1993) have supposedly been assessed, but there are a number of new species which have been described in recent years which have not been evaluated. These are a small fraction of the total number of mammal species, so for the purposes of the analysis, all species are said to have been assessed. The quality of the mammal assessments is highly variable, with many being based on relatively poor or sparse information in the case of the rodents, insectivores and insectivorous bats, although the status of many of the latter species was re-evaluated during the preparation of the bat action plan which has been submitted for publication.

Nomenclature

Where possible, standard world checklists have been used in order to promote nomenclatural stability. In a few instances Specialist Groups have used alternative systematic opinion and provided justification for doing so. All names of taxa on the Red List were checked and verified as far as was possible. In doing this, the correct authority name was included in an attempt to clarify what species concept is being followed. The names of the phyla and classes used in general follow Margulis and Schwartz (1988), but there are a number of deviations based on new evidence and thinking, particularly with regards to the plant groups. The following paragraphs note the main taxonomic sources used.

Mammals

The names of mammal orders, families and contents of families follow Wilson and Reeder (1993). Species nomenclature generally also follows this source, except when a Specialist Group has expressed a very strong preference for another system, or has used nomenclature different from Wilson and Reeder and we have been unable to resolve subsequent ambiguities about the population content of the species concerned and their distribution. Principal departures from Wilson and Reeder are relatively few in number, and are found in the primates and the bovids. The primates are undergoing a major taxonomic revision, which will appear in a book by Colin Groves that is soon to be published. Much of this revised taxonomy has been adopted by the SSC Primate Specialist Group and used in the 2000 Red List. The recent sixth edition of Walker's Mammals of the World (Novak 1999) proved to be very useful in clarifying various species concepts and for obtaining information for the documentation requirements.

Birds

Nomenclature for genera and families generally follows Sibley and Monroe (1990, 1993). Solely to maintain uniformity with *Threatened Birds of the World*, we use the names of orders and families, and the species content of families of Morony e*t al.* (1975). BirdLife International (2000) have used subfamily names to split up the larger families like the Muscicapidae, Embirizidae and Ploceidae in *Threatened Birds of the World*. These subfamilies are not used in the 2000 Red List.

Reptiles

Turtles and tortoises generally follow Iverson (1992); crocodilians follow King and Burke (1989); tuatara systematics are after Daugherty *et al.* (1990). Names in common use, including those used by Specialist Groups or in national sources, have been employed for other groups of reptiles. Increasing use is being made of *The EMBL Reptile Database* compiled by Peter Uetz (Uetz and Etzold 1996), and made available on the World Wide Web at: http://www.embl-heidelberg.de/~uetz/LivingReptiles.html. This is rapidly becoming the standard global checklist for reptiles.

Amphibians

Nomenclature generally follows Frost (1985) as updated by Duellman (1993). The Amphibian Species of the World Database is now available on the World Wide Web and is updated regularly, so this has become the source for any recent changes: http://research.amnh.org/cgi-bin/herpetology/amphibia. Another important web site for documentation on amphibian species, especially those in decline is the Amphibia Web Database at http://elib.cs.berkeley.edu/aw/.

Fishes

The names of orders, families, and the species content of families currently follows Eschmeyer (1990), but a number have been updated to be in line with the new thinking presented in Eschmeyer (1998). Some of the fish names used are derived from national sources or from Specialist Groups. Extensive recent taxonomic changes mean that the status of many species on the Red List needs to be re-assessed. This was not possible for the 2000 Red List, and the names and assessments are left as they appeared in the 1996 Red List, but it is important that this issue be resolved soon. An updated version of Eschmeyer's work is maintained as part of the comprehensive ICLARM (International Centre for Living Aquatic Resources Management) database (FishBase) which is available through the Species 2000 web site: http://www.sp2000.org/.

Invertebrates

Parker (1982) has generally been followed for nomenclature at class, order and family level. There is a lack of widely accepted class-level checklists for invertebrates and in the absence of such sources no attempt has been made to standardize names for inclusion. The Integrated Taxonomic Information Service (ITIS) web site developed jointly by the US Departments of Agriculture and US Geological Survey, is a useful source for a number of global and North American checklists covering a wide range of taxonomic groups including many invertebrates: http:/www.itis.usda.gov/.

Plants

For plant families and genera Brummitt (1992) is generally followed, but for the content of genera reference is made to a wide range of taxonomic treatments including papers on individual species, monographic treatments, standard floras, global checklists (e.g. Farjon 1998) and even site-specific checklists (e.g. Cable and Cheek 1998). The taxonomy of plant families and orders is undergoing major revision at present. Until such time that some level of stability is achieved, the orders of Cronquist (1981, 1988) are followed. Specific names are frequently checked against the International Plant Names Index (http://www.ipni.org/) which incorporates *Index Kewensis*, the Gray Index and the Australian Plant Names Index. The author citations for species follow Brummitt and Powell (1992) and as updated on the IPNI web site.

Undescribed species

Undescribed species are accepted on the Red List only under the following conditions:

- There is general agreement that the undescribed taxon is in fact a good species.
- Clear distribution information can be provided.
- Listing the undescribed species will potentially aid in its conservation.
- Specimen reference numbers (voucher collection details) are provided by which the species can be traced without confusion.
- The museum, herbarium or other institution holding the collection and the individual responsible for the proposal can be identified.
- Whenever possible a common name can be added.

Undescribed species are represented in the 2000 Red List by the generic name and the abbreviation sp. or sp. nov. Details of specimen numbers and institution should ideally be included in parentheses after the sp. nov. There are some instances where this has been done, but in many cases there have been requests for this information to be withheld. The Red List Programme Office should be contacted if further details are required.

Annex 4. Habitat types authority file

The habitat types listed here are those used in the documentation of species on the 2000 Red List. There is no single globally accepted habitat classification system currently available. The team developing the Species Information Service therefore suggested that the Global Land Cover Characterization system from the US Geological Service Earth Resources Observation Systems (EROS) Data Center (http://edcdaac.usgs.gov/glcc/glcc.html) be used. This system has the advantage of providing a uniform global classification system, which is geo-spatially explicit. The categories shown here are higher level amalgamations of many categories within the GLCC system, in an effort to make it simpler and easier to use. The end result, however, was far from ideal and it proved extremely difficult to interpolate habitat descriptions from the literature to an appropriate category in this list. In addition the GLCC system is largely focussed on terrestrial systems so freshwater and marine habitats are poorly classified, if at all. A separate field in the database for habitat notes allowed additional habitats to be recorded and from these some new habitats have been added to the list. The habitat list is currently under review.

Arable agriculture – cereals

Arable agriculture – excluding cereals

Bamboo

Beaches and dunes

Coastal rocky cliffs and slopes

Coastlines/lagoons/estuaries

Compounded coastlines
(beaches and rocky cliffs mixed)

Coniferous forest

Continental shelf waters

Coral reefs

Crop – grass and shrub mixture

Crops and urban

Crops and water mixtures
(including irrigated cropland)

Deep sea – Oceanic

Desert

Eucalyptus

Freshwater lakes/ponds/dams

Glacier ice

Grasses and shrub mosaic

Grassland

Heath scrub (cool)

Lowland tropical rainforest

Mangroves

Mediterranean scrub

Montane tropical rainforest

Open sea

Plantations

Polar and alpine bare soils

Rivers and streams

Rocks

Saline lakes/ponds/dams

Salt pans and playas

Seagrass beds

Semi-desert

Shrublands

Succulent and thorn scrub

Swamps/marshes/bogs

Temperate broadleaf forest

Temperate forest and field mosaics

Temperate mixed forest (coniferous and broadleaf)

Tropical degraded forest

Tropical monsoon and dry forest

Tropical savannah woodland
(with grass dominated understorey)

Tundra

Urban

Wooded tundra

Unknown/ Unspecified

Annex 5. Threat types authority file

The hierarchical structure of the major threat types used in the documentation of the 2000 Red List is shown here. The SSC's Species Information Service in consultation with the SSC network and key partners like BirdLife International developed these categories of threat in order to investigate major threatening processes..

1. Habitat Loss (primarily human induced)
    1.1. Agriculture
        1.1.1. Arable farming/horticulture
        1.1.2. Small-holder farming
        1.1.3. Shifting agriculture
        1.1.4. Livestock ranching
        1.1.5. Grazing
        1.1.6. Timber plantations
        1.1.7. Crop plantations
        1.1.8. Aquaculture
        1.1.9. Other
    1.2. Extraction
        1.2.1. Mining
        1.2.2. Fisheries
            1.2.3 Timber
            1.2.3.1. Clear-cutting
            1.2.3.2. Selective logging
            1.2.3.3. Firewood and charcoal production
        1.2.4. Harvesting – non-woody vegetation
        1.2.5. Mangrove removal
        1.2.6. Coral reef removal
        1.2.7. Groundwater extraction
        1.2.8. Other
    1.3. Development
        1.3.1. Industry
        1.3.2. Human settlement
        1.3.3. Tourism
        1.3.4. Infrastructure (roads, dams, power lines, etc.)
        1.3.5. Other
    1.4. Unspecified causes
        1.4.1. Fragmentation
        1.4.2. Deforestation
        1.4.3. Drainage/filling in of wetlands/coastlines
        1.4.4. Replacement by ground waste
        1.4.5. Soil loss/erosion
        1.4.6. Deliberate fires
        1.4.7. Other
2. Direct Loss/Exploitation
    2.1. Hunting and collecting
        2.1.1. Food
        2.1.2. Sport
        2.1.3. Cultural Use
        2.1.4. Traditional medicine
        2.1.5. Persecution
            2.1.5.1. Intentional poisoning (control)
        2.1.6. Other
    2.2. Trade
        2.2.1. Legal
            2.2.1.1. Food
            2.2.1.2. Commodities
            2.2.1.3. Traditional medicine
            2.2.1.4. Other
        2.2.2. Illegal
            2.2.2.1. Food
            2.2.2.2. Commodities
            2.2.2.3. Traditional medicine
            2.2.2.4. Other
        2.2.3. Legality unknown
            2.2.3.1. Food
            2.2.3.2. Commodities
            2.2.3.3. Traditional medicine
            2.2.3.4. Other
    2.3. Accidental mortality
        2.3.1. Trapping
        2.3.2. Hooking
        2.3.3. Netting
        2.3.4. Dynamite/explosives
        2.3.5. Poisoning
        2.3.6. Entanglement
        2.3.7. Pylon collision
        2.3.8. Air strikes
        2.3.9. Other

- 3. Indirect Effects
  - 3.1. Human caused
    - 3.1.1. Recreation/tourism
    - 3.1.2. Research
    - 3.1.3. Deliberate fires
    - 3.1.4. Other
  - 3.2. Alien invasive species
    - 3.2.1. Competitors
    - 3.2.2. Predators
    - 3.2.3. Hybridizers
    - 3.2.4. Pathogens/parasites
    - 3.2.5. Habitat loss
    - 3.2.6. Other
  - 3.3. Ecological imbalance (changes in native species dynamics)
    - 3.3.1. Competitors
    - 3.3.2. Predators
    - 3.3.3. Hybridizers
    - 3.3.4. Pathogens/parasites
    - 3.3.5. Habitat loss
    - 3.3.6. Loss of prey base
    - 3.3.7. Lack of pollinators
    - 3.3.8. Other
- 4. Natural disasters
  - 4.1. Volcanoes
  - 4.2. Drought
  - 4.3. Wildfire
  - 4.4. Storms/flooding
  - 4.5. Other
- 5. Atmospheric pollution
  - 5.1. Global warming/oceanic warming
  - 5.2. Acid precipitation
  - 5.3. Ozone hole effects
  - 5.4. Other
- 6. Land/Water pollution
  - 6.1. Pesticides/chemical pollution
  - 6.2. Industrial pollution
  - 6.3. Oil slicks
  - 6.4. Other
- 7. Intrinsic Factors
  - 7.1. Poor dispersal
  - 7.2. Poor recruitment/reproduction/regeneration
  - 7.3. High juvenile mortality
  - 7.4. Inbreeding
  - 7.5. Other
- 8. Other
- 9. Unknown

Annex 6. The 1994 IUCN Red List categories and criteria

Prepared by the IUCN Species Survival Commission
As approved by the 40th Meeting of the IUCN Council Gland, Switzerland, 30 November 1994

I. Introduction

1. The threatened species categories now used in Red Data Books and Red Lists have been in place, with some modification, for almost 30 years. Since their introduction these categories have become widely recognized internationally, and they are now used in a whole range of publications and listings, produced by IUCN as well as by numerous governmental and non-governmental organizations. The Red Data Book categories provide an easily and widely understood method for highlighting those species under higher extinction risk, so as to focus attention on conservation measures designed to protect them.

2. The need to revise the categories has been recognized for some time. In 1984, the SSC held a symposium, 'The Road to Extinction' (Fitter and Fitter 1987), which examined the issues in some detail, and at which a number of options were considered for the revised system. However, no single proposal resulted. The current phase of development began in 1989 with a request from the SSC Steering Committee to develop a new approach that would provide the conservation community with useful information for action planning.

In this document, proposals for new definitions for Red List categories are presented. The general aim of the new system is to provide an explicit, objective framework for the classification of species according to their extinction risk.

The revision has several specific aims:

- to provide a system that can be applied consistently by different people;
- to improve the objectivity by providing those using the criteria with clear guidance on how to evaluate different factors which affect risk of extinction;
- to provide a system which will facilitate comparisons across widely different taxa;
- to give people using threatened species lists a better understanding of how individual species were classified.

3. The proposals presented in this document result from a continuing process of drafting, consultation and validation. It was clear that the production of a large number of draft proposals led to some confusion, especially as each draft has been used for classifying some set of species for conservation purposes. To clarify matters, and to open the way for modifications as and when they became necessary, a system for version numbering was applied as follows:

Version 1.0: Mace and Lande (1991)
The first paper discussing a new basis for the categories, and presenting numerical criteria especially relevant for large vertebrates.

Version 2.0: Mace *et al.* (1992)
A major revision of Version 1.0, including numerical criteria appropriate to all organisms and introducing the non-threatened categories.

Version 2.1: IUCN (1993)
Following an extensive consultation process within SSC, a number of changes were made to the details of the criteria, and fuller explanation of basic principles was included. A more explicit structure clarified the significance of the non-threatened categories.

Version 2.2: Mace and Stuart (1994)
Following further comments received and additional validation exercises, some minor changes to the criteria were made. In addition, the Susceptible category present in Versions 2.0 and 2.1 was subsumed into the Vulnerable category. A precautionary application of the system was emphasized.

Final Version
This final document, which incorporates changes as a result of comments from IUCN members, was adopted by the IUCN Council in December 1994.

All future taxon lists including categorizations should be based on this version, and not the previous ones.

4. In the rest of this document the proposed system is outlined in several sections. The Preamble presents some basic information about the context and structure of the proposal, and the procedures that are to be followed in applying the definitions to species. This is followed by a section giving definitions of terms used. Finally the definitions are presented, followed by the quantitative criteria used for classification within the threatened categories. It is important for the effective functioning of the new system that all sections are read and understood, and the guidelines followed.

References:

Fitter, R., and M. Fitter, eds. (1987) *The Road to Extinction*. IUCN, Gland, Switzerland.

IUCN. (1993) Draft IUCN Red List Categories. IUCN, Gland, Switzerland.

Mace, G. M. *et al.* (1992) The development of new criteria for listing species on the IUCN Red List. *Species* **19**: 16–22.

Mace, G. M. and R. Lande. (1991) Assessing extinction threats: toward a re-evaluation of IUCN threatened species categories. *Conserv. Biol.* **5.2**: 148–157.

Mace, G. M. and S. N. Stuart. (1994) Draft IUCN Red List Categories, Version 2.2. *Species* **21–22**: 13–24.

II. Preamble

The following points present important information on the use and interpretation of the categories (= Critically Endangered, Endangered, etc.), criteria (= A to E), and sub-criteria (= a,b etc., i,ii etc.):

1. Taxonomic level and scope of the categorization process

The criteria can be applied to any taxonomic unit at or below the species level. The term 'taxon' in the following notes, definitions and criteria is used for convenience, and may represent species or lower taxonomic levels, including forms that are not yet formally described. There is sufficient range among the different criteria to enable the appropriate listing of taxa from the complete taxonomic spectrum, with the exception of micro-organisms. The criteria may also be applied within any specified geographical or political area although in such cases special notice should be taken of point 11 below. In presenting the results of applying the criteria, the taxonomic unit and area under consideration should be made explicit. The categorization process should only be applied to wild populations inside their natural range, and to populations resulting from benign introductions (defined in the draft IUCN Guidelines for Re-introductions as "..an attempt to establish a species, for the purpose of conservation, outside its recorded distribution, but within an appropriate habitat and ecogeographical area").

2. Nature of the categories

All taxa listed as Critically Endangered qualify for Vulnerable and Endangered, and all listed as Endangered qualify for Vulnerable. Together these categories are described as 'threatened'. The threatened species categories form a part of the overall scheme. It will be possible to place all taxa into one of the categories (see Figure A6.1).

3. Role of the different criteria

For listing as Critically Endangered, Endangered or Vulnerable there is a range of quantitative criteria; meeting any one of these criteria qualifies a taxon for listing at that level of threat. Each species should be evaluated against all the criteria. The different criteria (A-E) are derived from a wide review aimed at detecting risk factors across the broad range of organisms and the diverse life histories they exhibit. Even though some criteria will be inappropriate for

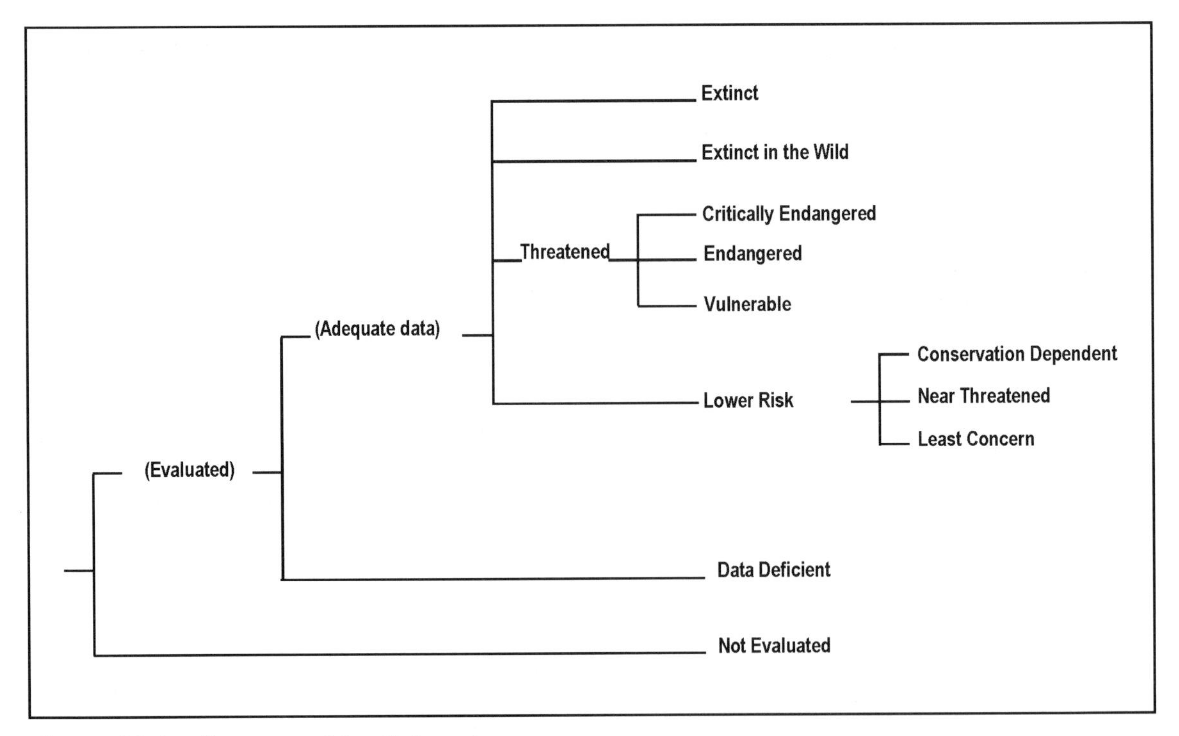

Figure A6.1. Structure of the Categories

certain taxa (some taxa will never qualify under these however close to extinction they come), there should be criteria appropriate for assessing threat levels for any taxon (other than micro-organisms). The relevant factor is whether any one criterion is met, not whether all are appropriate or all are met. Because it will never be clear which criteria are appropriate for a particular species in advance, each species should be evaluated against all the criteria, and any criterion met should be listed.

4. Derivation of quantitative criteria

The quantitative values presented in the various criteria associated with threatened categories were developed through wide consultation and they are set at what are generally judged to be appropriate levels, even if no formal justification for these values exists. The levels for different criteria within categories were set independently but against a common standard. Some broad consistency between them was sought. However, a given taxon should not be expected to meet all criteria (A–E) in a category; meeting any one criterion is sufficient for listing.

5. Implications of listing

Listing in the categories of Not Evaluated and Data Deficient indicates that no assessment of extinction risk has been made, though for different reasons. Until such time as an assessment is made, species listed in these categories should not be treated as if they were non-threatened, and it may be appropriate (especially for Data Deficient forms) to give them the same degree of protection as threatened taxa, at least until their status can be evaluated.

Extinction is assumed here to be a chance process. Thus, a listing in a higher extinction risk category implies a higher expectation of extinction, and over the time-frames specified more taxa listed in a higher category are expected to go extinct than in a lower one (without effective conservation action). However, the persistence of some taxa in high risk categories does not necessarily mean their initial assessment was inaccurate.

6. Data quality and the importance of inference and projection

The criteria are clearly quantitative in nature. However, the absence of high quality data should not deter attempts at applying the criteria, as methods involving estimation, inference and projection are emphasized to be acceptable throughout. Inference and projection may be based on extrapolation of current or potential threats into the future (including their rate of change), or of factors related to population abundance or distribution (including dependence on other taxa), so long as these can reasonably be supported. Suspected or inferred patterns in either the recent past, present or near future can be based on any of a series of related factors, and these factors should be specified.

Taxa at risk from threats posed by future events of low probability but with severe consequences (catastrophes) should be identified by the criteria (e.g. small distributions, few locations). Some threats need to be identified particularly early, and appropriate actions taken, because their effects are irreversible, or nearly so (pathogens, invasive organisms, hybridization).

7. Uncertainty

The criteria should be applied on the basis of the available evidence on taxon numbers, trend and distribution, making due allowance for statistical and other uncertainties. Given that data are rarely available for the whole range or population of a taxon, it may often be appropriate to use the information that is available to make intelligent inferences about the overall status of the taxon in question. In cases where a wide variation in estimates is found, it is legitimate to apply the precautionary principle and use the estimate (providing it is credible) that leads to listing in the category of highest risk.

Where data are insufficient to assign a category (including Lower Risk), the category of 'Data Deficient' may be assigned. However, it is important to recognize that this category indicates that data are inadequate to determine the degree of threat faced by a taxon, not necessarily that the taxon is poorly known. In cases where there are evident threats to a taxon through, for example, deterioration of its only known habitat, it is important to attempt threatened listing, even though there may be little direct information on the biological status of the taxon itself. The category 'Data Deficient' is not a threatened category, although it indicates a need to obtain more information on a taxon to determine the appropriate listing.

8. Conservation actions in the listing process

The criteria for the threatened categories are to be applied to a taxon whatever the level of conservation action affecting it. In cases where it is only conservation action that prevents the taxon from meeting the threatened criteria, the designation of 'Conservation Dependent' is appropriate. It is important to emphasize here that a taxon require conservation action even if it is not listed as threatened.

9. Documentation

All taxon lists including categorization resulting from these criteria should state the criteria and sub-criteria that were met. No listing can be accepted as valid unless at least one criterion is given. If more than one criterion or sub-criterion was met, then each should be listed. However, failure to mention a criterion should not necessarily imply that it was not met. Therefore, if a re-evaluation indicates that the documented criterion is no longer met, this should not result in automatic down-listing. Instead, the taxon should be re-evaluated with respect to all criteria to indicate its status. The factors responsible for triggering the criteria, especially where inference and projection are used, should at least be logged by the evaluator, even if they cannot be included in published lists.

10. Threats and priorities

The category of threat is not necessarily sufficient to determine priorities for conservation action. The category of threat simply provides an assessment of the likelihood of extinction under current circumstances, whereas a system for assessing priorities for action will include numerous other factors concerning conservation action such as costs, logistics, chances of success, and even perhaps the taxonomic distinctiveness of the subject.

11. Use at regional level

The criteria are most appropriately applied to whole taxa at a global scale, rather than to those units defined by regional or national boundaries. Regionally or nationally based threat categories, which are aimed at including taxa that are threatened at regional or national levels (but not necessarily throughout their global ranges), are best used with two key pieces of information: the global status category for the taxon, and the proportion of the global population or range that occurs within the region or nation. However, if applied at regional or national level it must be recognized that a global category of threat may not be the same as a regional or national category for a particular taxon. For example, taxa classified as Vulnerable on the basis of their global declines in numbers or range might be Lower Risk within a particular region where their populations are stable. Conversely, taxa classified as Lower Risk globally might be Critically Endangered within a particular region where numbers are very small or declining, perhaps only because they are at the margins of their global range. IUCN is still in the process of developing guidelines for the use of national red list categories.

12. Re-evaluation

Evaluation of taxa against the criteria should be carried out at appropriate intervals. This is especially important for taxa listed under Near Threatened, or Conservation Dependent, and for threatened species whose status is known or suspected to be deteriorating.

13. Transfer between categories

There are rules to govern the movement of taxa between categories. These are as follows: (A) A taxon may be moved from a category of higher threat to a category of lower threat if none of the criteria of the higher category has been met for five years or more. (B) If the original classification is found to have been erroneous, the taxon may be transferred to the appropriate category or removed from the threatened categories altogether, without delay (but see Section 9). (C) Transfer from categories of lower to higher risk should be made without delay.

14. Problems of scale

Classification based on the sizes of geographic ranges or the patterns of habitat occupancy is complicated by problems of spatial scale. The finer the scale at which the distributions or habitats of taxa are mapped, the smaller the area will be that they are found to occupy. Mapping at finer scales reveals more areas in which the taxon is unrecorded. It is impossible to provide any strict but general rules for mapping taxa or habitats; the most appropriate scale will depend on the taxa in question, and the origin and comprehensiveness of the distributional data. However, the thresholds for some criteria (e.g. Critically Endangered) necessitate mapping at a fine scale.

III. Definitions

1. Population

Population is defined as the total number of individuals of the taxon. For functional reasons, primarily owing to differences between life-forms, population numbers are expressed as numbers of mature individuals only. In the case of taxa obligately dependent on other taxa for all or part of their life cycles, biologically appropriate values for the host taxon should be used.

2. Subpopulations

Subpopulations are defined as geographically or otherwise distinct groups in the population between which there is little exchange (typically one successful migrant individual or gamete per year or less).

3. Mature individuals

The number of mature individuals is defined as the number of individuals known, estimated or inferred to be capable of reproduction. When estimating this quantity the following points should be borne in mind:

- Where the population is characterized by natural fluctuations the minimum number should be used.
- This measure is intended to count individuals capable of reproduction and should therefore exclude individuals that are environmentally, behaviourally or otherwise reproductively suppressed in the wild.
- In the case of populations with biased adult or breeding sex ratios it is appropriate to use lower estimates for the number of mature individuals which take this into account (e.g. the estimated effective population size).
- Reproducing units within a clone should be counted as individuals, except where such units are unable to survive alone (e.g. corals).
- In the case of taxa that naturally lose all or a subset of mature individuals at some point in their life cycle, the estimate should be made at the appropriate time, when mature individuals are available for breeding.

4. Generation

Generation may be measured as the average age of parents in the population. This is greater than the age at first breeding, except in taxa where individuals breed only once.

5. Continuing decline

A continuing decline is a recent, current or projected future decline whose causes are not known or not adequately controlled and so is liable to continue unless remedial measures are taken. Natural fluctuations will not normally count as a continuing decline, but an observed decline should not be considered to be part of a natural fluctuation unless there is evidence for this.

6. Reduction

A reduction (criterion A) is a decline in the number of mature individuals of at least the amount (%) stated over the time period (years) specified, although the decline need not still be continuing. A reduction should not be interpreted s part of a natural fluctuation unless there is good evidence for this. Downward trends that are part of natural fluctuations will not normally count as a reduction.

7. Extreme fluctuations

Extreme fluctuations occur in a number of taxa where population size or distribution area varies widely, rapidly and frequently, typically with a variation greater than one order of magnitude (i.e., a tenfold increase or decrease).

8. Severely fragmented

Severely fragmented refers to the situation where increased extinction risks to the taxon result from the fact that most individuals within a taxon are found in small and relatively isolated subpopulations. These small subpopulations may go extinct, with a reduced probability of recolonization.

9. Extent of occurrence

Extent of occurrence is defined as the area contained within the shortest continuous imaginary boundary which can be drawn to encompass all the known, inferred or projected sites of present occurrence of a taxon, excluding cases of vagrancy. This measure may exclude discontinuities or disjunctions within the overall distributions of taxa (e.g., large areas of obviously unsuitable habitat) (but see 'area of occupancy'). Extent of occurrence can often be measured by a minimum convex polygon (the smallest polygon in which no internal angle exceeds 180 degrees and which contains all the sites of occurrence).

10. Area of occupancy

Area of occupancy is defined as the area within its 'extent of occurrence' (see definition) which is occupied by a taxon, excluding cases of vagrancy. The measure reflects the fact that a taxon will not usually occur throughout the area of its extent of occurrence, which may, for example, contain unsuitable habitats. The area of occupancy is the smallest area essential at any stage to the survival of existing populations of a taxon (e.g. colonial nesting sites, feeding sites for migratory taxa). The size of the area of occupancy will be a function of the scale at which it is measured, and should be at a scale appropriate to relevant biological aspects of the taxon. The criteria include values in km^2, and thus to avoid errors in classification, the area of occupancy should be measured on grid squares (or equivalents) which are sufficiently small (see Figure A6.2).

11. Location

Location defines a geographically or ecologically distinct area in which a single event (e.g. pollution) will soon affect all individuals of the taxon present. A location usually, but not always, contains all or part of a subpopulation of the taxon, and is typically a small proportion of the taxon's total distribution.

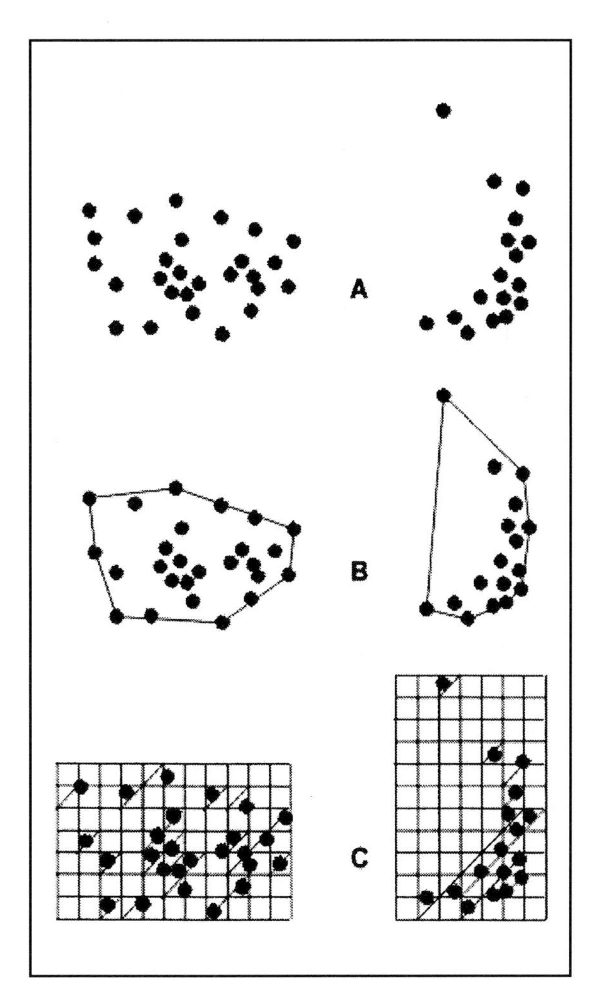

Figure A6.2. Two examples of the distinction between extent of occurrence and area of occupancy. A is the spatial distribution of known, inferred or projected sites of occurrence. B shows one possible boundary to the extent of occurrence, which is the measured area within this boundary. C shows one measure of area of occupancy which can be measured by the sum of the occupied grid squares.

12. Quantitative analysis

A quantitative analysis is defined here as the technique of population viability analysis (PVA), or any other quantitative form of analysis, which estimates the extinction probability of a taxon or population based on the known life history and specified management or non-management options. In presenting the results of quantitative analyses the structural equations and the data should be explicit.

Two examples of the distinction between extent of occurrence and area of occupancy. (a) is the spatial distribution of known, inferred or projected sites of occurrence. (b) shows one possible boundary to the extent of occurrence, which is the measured area within this boundary. (c) shows one measure of area of occupancy which can be measured by the sum of the occupied grid squares.

IV. The categories[1]

EXTINCT (EX)

A taxon is Extinct when there is no reasonable doubt that the last individual has died.

EXTINCT IN THE WILD (EW)

A taxon is Extinct in the Wild when it is known only to survive in cultivation, in captivity or as a naturalized population (or populations) well outside the past range. A taxon is presumed extinct in the wild when exhaustive surveys in known and/or expected habitat, at appropriate times (diurnal, seasonal, annual), throughout its historic range have failed to record an individual. Surveys should be over a time frame appropriate to the taxon's life cycle and life form.

CRITICALLY ENDANGERED (CR)

A taxon is Critically Endangered when it is facing an extremely high risk of extinction in the wild in the immediate future, as defined by any of the criteria (A to E) on pages 16 and 17.

ENDANGERED (EN)

A taxon is Endangered when it is not Critically Endangered but is facing a very high risk of extinction in the wild in the near future, as defined by any of the criteria (A to E) on pages 18 and 19.

VULNERABLE (VU)

A taxon is Vulnerable when it is not Critically Endangered or Endangered but is facing a high risk of extinction in the wild in the medium-term future, as defined by any of the criteria (A to E) on pages 20, 21 and 22.

LOWER RISK (LR)

A taxon is Lower Risk when it has been evaluated, does not satisfy the criteria for any of the categories Critically Endangered, Endangered or Vulnerable. Taxa included in the Lower Risk category can be separated into three subcategories:

- **Conservation Dependent (cd).** Taxa which are the focus of a continuing taxon-specific or habitat-specific conservation programme targeted towards the taxon in question, the cessation of which would result in the taxon qualifying for one of the threatened categories above within a period of five years.
- **Near Threatened (nt)**. Taxa which do not qualify for Conservation Dependent, but which are close to qualifying for Vulnerable.
- **Least Concern (lc).** Taxa which do not qualify for Conservation Dependent or Near Threatened.

DATA DEFICIENT (DD)

A taxon is Data Deficient when there is inadequate information to make a direct, or indirect, assessment of its risk of extinction based on its distribution and/or population status. A taxon in this category may be well studied, and its biology well known, but appropriate data on abundance and/or distribution is lacking. Data Deficient is therefore not a category of threat or Lower Risk. Listing of taxa in this category indicates that more information is required and acknowledges the possibility that future research will show that threatened classification is appropriate. It is important to make positive use of whatever data are available. In many cases great care should be exercized in choosing between DD and threatened status. If the range of a taxon is suspected to be relatively circumscribed, if a considerable period of time has elapsed since the last record of the taxon, threatened status may well be justified.

NOT EVALUATED (NE)

A taxon is Not Evaluated when it is has not yet been assessed against the criteria.

V. The Criteria for Critically Endangered, Endangered and Vulnerable

CRITICALLY ENDANGERED (CR)

A taxon is Critically Endangered when it is facing an extremely high risk of extinction in the wild in the immediate future, as defined by any of the following criteria (A to E):

1 As in previous IUCN categories, the abbreviation of each category (in parenthesis) follows the English denominations when translated into other languages. The page numbers referred to in this document are those in the official printed IUCN Red List Categories booklet, copies of which are available on request from IUCN (address on inside cover of this publication).

(A) Population reduction in the form of either of the following:

(1) An observed, estimated, inferred or suspected reduction of at least 80% over the last 10 years or three generations, whichever is the longer, based on (and specifying) any of the following:

(a) direct observation

(b) an index of abundance appropriate for the taxon

(c) a decline in area of occupancy, extent of occurrence and/or quality of habitat

(d) actual or potential levels of exploitation

(e) the effects of introduced taxa, hybridization, pathogens, pollutants, competitors or parasites.

(2) A reduction of at least 80%, projected or suspected to be met within the next 10 years or three generations, whichever is the longer, based on (and specifying) any of (b), (c), (d) or (e) above.

(B) Extent of occurrence estimated to be less than 100km^2 or area of occupancy estimated to be less than 10km^2, and estimates indicating any two of the following:

(1) Severely fragmented or known to exist at only a single location.

(2) Continuing decline, observed, inferred or projected, in any of the following:

(a) extent of occurrence

(b) area of occupancy

(c) area, extent and/or quality of habitat

(d) number of locations or subpopulations

(e) number of mature individuals.

(3) Extreme fluctuations in any of the following:

(a) extent of occurrence

(b) area of occupancy

(c) number of locations or subpopulations

(d) number of mature individuals.

(C) Population estimated to number less than 250 mature individuals and either:

(1) An estimated continuing decline of at least 25% within three years or one generation, whichever is longer or

(2) A continuing decline, observed, projected, or inferred, in numbers of mature individuals and population structure in the form of either:

(a) severely fragmented (i.e. no subpopulation estimated to contain more than 50 mature individuals)

(b) all individuals are in a single subpopulation.

(D) Population estimated to number less than 50 mature individuals.

(E) Quantitative analysis showing the probability of extinction in the wild is at least 50% within 10 years or three generations, whichever is the longer.

ENDANGERED (EN)

A taxon is Endangered when it is not Critically Endangered but is facing a very high risk of extinction in the wild in the near future, as defined by any of the following criteria (A to E):

(A) Population reduction in the form of either of the following:

(1) An observed, estimated, inferred or suspected reduction of at least 50% over the last 10 years or three generations, whichever is the longer, based on (and specifying) any of the following:

(a) direct observation

(b) an index of abundance appropriate for the taxon

(c) a decline in area of occupancy, extent of occurrence and/or quality of habitat

(d) actual or potential levels of exploitation

(e) the effects of introduced taxa, hybridization, pathogens, pollutants, competitors or parasites.

(2) A reduction of at least 50%, projected or suspected to be met within the next 10 years or three generations, whichever is the longer, based on (and specifying) any of (b), (c), (d), or (e) above.

(B) Extent of occurrence estimated to be less than 5000 km^2 or area of occupancy estimated to be less than 500 km^2, and estimates indicating any two of the following:

(1) Severely fragmented or known to exist at no more than five locations.

(2) Continuing decline, inferred, observed or projected, in any of the following:

(a) extent of occurrence

(b) area of occupancy

(c) area, extent and/or quality of habitat

(d) number of locations or subpopulations

(e) number of mature individuals.

(3) Extreme fluctuations in any of the following:

(a) extent of occurrence

(b) area of occupancy

(c) number of locations or subpopulations

(d) number of mature individuals.

(C) Population estimated to number less than 2500 mature individuals and either:

(1) An estimated continuing decline of at least 20% within five years or two generations, whichever is longer, or

(2) A continuing decline, observed, projected, or inferred, in numbers of mature individuals and population structure in the form of either:

(a) severely fragmented (i.e. no subpopulation estimated to contain more than 250 mature individuals)

(b) all individuals are in a single subpopulation.

(D) Population estimated to number less than 250 mature individuals.

(E) Quantitative analysis showing the probability of extinction in the wild is at least 20% within 20 years or five generations, whichever is the longer.

VULNERABLE (VU)

A taxon is Vulnerable when it is not Critically Endangered or Endangered but is facing a high risk of extinction in the wild in the medium-term future, as defined by any of the following criteria (A to E):

(A) Population reduction in the form of either of the following:

(1) An observed, estimated, inferred or suspected reduction of at least 20% over the last 10 years or three generations, whichever is the longer, based on (and specifying) any of the following:

(a) direct observation

(b) an index of abundance appropriate for the taxon

(c) a decline in area of occupancy, extent of occurrence and/or quality of habitat

(d) actual or potential levels of exploitation

(e) the effects of introduced taxa, hybridization, pathogens, pollutants, competitors or parasites.

(2) A reduction of at least 20%, projected or suspected to be met within the next ten years or three generations, whichever is the longer, based on (and specifying) any of (b), (c), (d) or (e) above.

(B) Extent of occurrence estimated to be less than 20,000 km^2 or area of occupancy estimated to be less than 2000 km^2, and estimates indicating any two of the following:

(1) Severely fragmented or known to exist at no more than ten locations.

(2) Continuing decline, inferred, observed or projected, in any of the following:

(a) extent of occurrence

(b) area of occupancy

(c) area, extent and/or quality of habitat

(d) number of locations or subpopulations

(e) number of mature individuals

(3) Extreme fluctuations in any of the following:

(a) extent of occurrence

(b) area of occupancy

(c) number of locations or subpopulations

(d) number of mature individuals

(C) Population estimated to number less than 10,000 mature individuals and either:

(1) An estimated continuing decline of at least 10% within 10 years or three generations, whichever is longer, or

(2) A continuing decline, observed, projected, or inferred, in numbers of mature individuals and population structure in the form of either:

(a) severely fragmented (i.e. no subpopulation estimated to contain more than 1000 mature individuals)

(b) all individuals are in a single subpopulation

(D) Population very small or restricted in the form of either of the following:

(1) Population estimated to number less than 1000 mature individuals.

(2) Population is characterized by an acute restriction in its area of occupancy (typically less than 100 km^2) or in the number of locations (typically less than five). Such a taxon would thus be prone to the effects of human activities (or stochastic events whose impact is increased by human activities) within a very short period of time in an unforeseeable future, and is thus capable of becoming Critically Endangered or even Extinct in a very short period.

(E) Quantitative analysis showing the probability of extinction in the wild is at least 10% within 100 years.

Annex 7. Summary of the results of the review of IUCN Red List categories and criteria 1996–2000

by Georgina M. Mace

Background to the criteria review

In 1994, IUCN adopted new criteria for assessing extinction risks to species, published in IUCN Red Lists. About 15,000 species were assessed using the new criteria for the *1996 IUCN Red List of Threatened Animals*. Of these, 5205 were listed as threatened with extinction. The relative objectivity of the new listings has made them an excellent tool for observing changes in status over time and this new method has attracted great interest from wildlife agencies and management authorities, as well as the media. Not surprisingly, there were also controversial elements in the new publication, including fisheries species, long-lived species such as elephants and marine turtles, and the status of some small and very narrowly distributed endemic molluscs and invertebrates.

At the World Conservation Congress (WCC) in Montreal in October 1996, SSC was mandated under WCC Resolution 1.4 to:

> *"within available resources, urgently to complete its review of the IUCN Red List Categories and Criteria, in an open and transparent manner, in consultation with relevant experts, to ensure the criteria are effective indicators of risk of extinction across the broadest possible range of taxonomic categories, especially in relation to:*
>
> - *marine species, particularly fish, taking into account the dynamic nature of marine ecosystems;*
> - *species under management programmes;*
> - *the time periods over which declines are measured."*

Under the auspices of the IUCN/SSC Red List Programme, SSC set up a Criteria Review Working Group. The task of this group was to respond to the mandate given to SSC at the World Conservation Congress and to report back to the SSC Executive Committee.

The process

The Criteria Review Working Group consisted of 25 members, representing a wide range of expertise on animal and plant taxa, and including people with technical knowledge about extinction risk assessment, as well as experience in applying the Red List criteria. This Group has overseen the review and the final recommendations.

The review has been conducted in stages as outlined below.

Dates	Activity
Jan – Dec 1997	Correspondence and seeking input from the members of IUCN and SSC.
Jan – Feb 1998	Planning for Scoping Workshop.
March 1998	**Scoping Workshop**, London, UK. Funded by IUCN.
March – Sept. 1998	Planning and fund-raising for activities outlined by the Scoping Workshop.
October 1998	**Regional Assessment Working Group** Montreal, Canada. The meeting contributes views on regional assessments. Funding support from Canadian Wildlife Service.
January 1999	**Marine Workshop**. Tokyo, Japan. Funding from German Government. Evaluates issues related to marine species. Additional input from Japanese meeting on Risk Assessment.
May 1999	**Range Size, Habitat Areas and Dealing with Uncertainty Workshop**. Manly, Sydney, Australia. Funding from environment and technical agencies in New South Wales, Australia.
June 1999	**Criterion A Workshop.** Cambridge, UK. Funding from Finnish Government.
July 1999	**Review Workshop**. Cambridge, UK. Criteria Review Working Group meets to discuss recommendations from all workshop reports, and provide final set of recommendations. Funding from Finnish Government
September 1999	**Publication in *Species***. Draft of revised criteria prepared and published in *Species* for circulation to all SSC members and circulated to all IUCN members.
Sept – Nov 1999	Correspondence and seeking input from the members of IUCN and SSC.
December 1999	Submission of re-drafted proposals to SSC Executive

January 2000	Workshop to resolve geographical scale issues. Uppsala. Sweden. Funding from three Swedish agencies.
February 2000	Submission of revised IUCN Red List Categories and Criteria to IUCN Council for approval

The workshops from January to July 1999 followed directly from specific issues outlined by the Scoping Workshop in March 1998. Participants at these workshops were selected to reflect technical and practical expertise in the areas being discussed. All workshops addressed specific issues and attempted to deliver recommended courses of action through analysis and discussion. In order to provide continuity and coherence to the process, at least 4–5 members of the Criteria Review Working Group attended each topic-based workshop. In addition, each member of the group was requested to attend at least one of the workshops.

Written reports on the workshops provide all the supporting arguments and documentation for the final outcome of the review as presented here. All the workshop reports adhere to a common standard, are comprehensive and will be available as a package along with the final report from the Criteria Working Group. Copies of reports produced so far are available via the IUCN web site (http://www.iucn.org/themes/ssc/siteindx.htm) or they can be ordered directly from the IUCN Red List Programme Officer. A full outline of the draft proposals for amending the criteria was published in *Species* 31–32, pages 43–57.

Over 60 people have been involved directly in the workshops. All review group members have participated in at least one of the topical workshops. As a result of the review process, several new topics have become the focus of active research and publication in the academic community, e.g. handling uncertainty, scale and area measurement, life history impacts for threat status and the nature of declining populations.

Changes to the categories and criteria

The changes described in the section follow the sequence in the IUCN rules (see Annex 6).

Introduction

There is a need for a more explicit account of the role and purpose of the Red List (including the background and history to the current listing procedure). This should include an account of how a listing status should be interpreted, the relationship of the criteria to one another, their background in theoretical biology, and what they are and are not intended to indicate. The difference between measuring threats and assessing conservation priorities also needs to be expanded, as there are many people who interpret the Red List as a means of priority setting. The introduction was identified as one place where some of these issues should be dealt with in more detail; the remainder will be covered in the detailed user guidelines.

Outcome

- A new introduction explains the role and appropriate uses for the categories and criteria. New version numbering is added.

Preamble

Various changes were made to the Preamble to reflect changes elsewhere in the document. An area of particular importance concerned the handling of uncertainty in the criteria. Despite the fact that the notes accompanying the current criteria recognize the problem of data uncertainty, there is no clear guidance on how to deal with it in either the assessment of species or the interpretation of listings. This is an important problem that limits the use and interpretation of the Red List Criteria and Categories, and leads to irresolvable debates over particular issues. Many other problems with the criteria are related to this issue, e.g. the use of Data Deficient, the lack of criteria for Near Threatened, and the assessment of species whose status is known only from one small part of its range. New methods and approaches developed during the review provide a better understanding of uncertainty and offer a way forward.

Outcome

- Re-ordering of points for clarity.
- A few small editorial changes for clarity.
- A new Figure 1 to reflect changes to categories (later).
- New section on uncertainty with addition of detailed Annex 1 which provides full guidance on dealing with uncertainty that is consistent with the methods implemented in the RAMAS® Red List software package.
- A new section on regional level assessments, which refers to the guidelines produced by the Regional Applications Working Group.

Definitions

Many small changes were suggested in the review to improve clarity, consistency and/or accuracy in the definitions of terms used in the criteria.

Outcome

- Slightly revised versions of most definitions.
- New section to deal with scale problems under Area of Occupancy.
- New wording for quantitative analysis to ensure that its use is clear for cases where the modeling is of environmental rather than population processes and is not directly equivalent to applying PVA.

The Categories

Qualitative definitions

The qualitative definitions for the threatened categories tend to overstate the predictive accuracy of the system. They also do not adequately convey to the general reader the fact that it is the criteria that determine listing in the threatened categories and that this evaluation requires a scientifically based assessment. The difficulty is how to phrase them without using quantitative terms but still convey a sense of urgency.

Outcome

- New wording for qualitative definitions for threatened categories.

Conservation dependent

The current use of Conservation Dependent as an independent Red List Category is not logically consistent as a taxon can be both threatened and conservation dependent. In addition assessors have used this category in a variety of contexts making it less useful than was hoped. More logically Conservation Dependent could be used as a flag under all the threatened categories but this is not a satisfactory solution as it would require many difficult judgements to be made about the effectiveness of conservation programmes.

Outcome

- Removal of the category 'Conservation Dependent'

Near threatened

This category is increasingly being used more formally than was intended. At present it is very loosely defined, so better guidance is required on when and how to use it. The development of criteria has been suggested, but this option would create many difficulties. The guidelines will provide practical and more consistent methods for determining when a species should be listed as Near Threatened. This might be where a taxon meets only some sub-criteria or where there is a plausible assessment of a threatened category but the assessment based on best estimates leads to Least Concern. In addition this category would include some taxa that previously would have been listed as Conservation Dependent.

Outcome

- New definition for Near Threatened that is more specific about when it should be used and includes the species previously classified as Conservation Dependent.

Least Concern

This category was provided to differentiate species that had been evaluated, and found not to be threatened. This gives the impression that one is required to conduct a formal assessment for blatantly common (weedy) taxa. From basic observations it can be easily seen that most of these extremely common taxa would not qualify for listing even though they have not been put through a formal assessment.

Outcome

- New definition that makes its role clearer. The means that the old category of Lower Risk is no longer necessary.
- The changes to the categories resulted in a new figure for the structure of the IUCN Red List categories which is simplified compared to the 1994 version (see Figure A7.1).

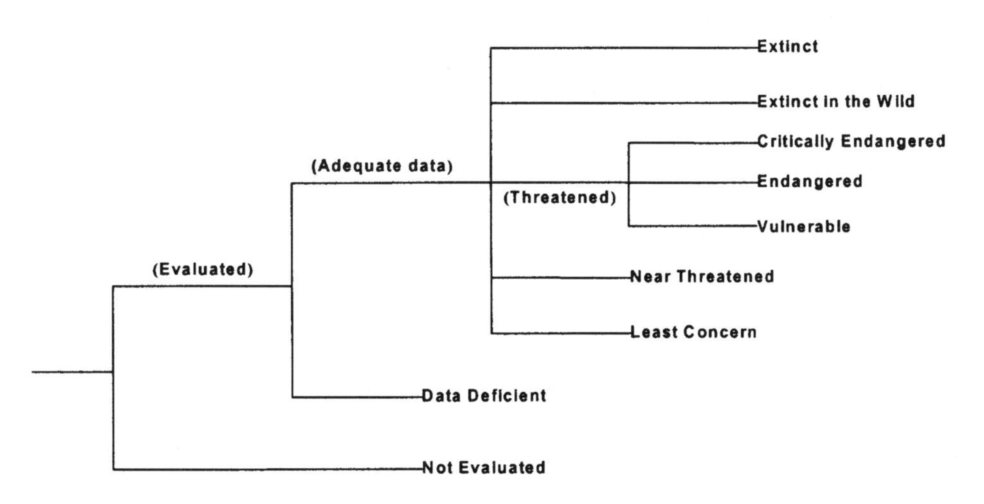

Figure A7.1. New structure for the IUCN categories and criteria

Changes to the Criteria

Criterion A

A number of problems with Criterion A were identified during the review process. The current quantitative thresholds, especially for Vulnerable may be too low. In addition, the rates of decline do not take into account managed populations that are being harvested down to levels at which higher yield is attained, or dramatic declines that occurred in the distant past but are now halted or even reversed. The criterion also does not provide guidance on projecting into the future, especially for long-lived species, where such assessments may be both unreliable and irrelevant. Greater clarity is also required on whether the criterion allows the use of a shifting time window for species where only a small amount of data is available. The confidence limits on declining population data are also an important issue as strict application of the precautionary principle could lead to over-listing under this criterion.

Outcome

- New subcriterion to provide higher decline rate thresholds for species that have ceased declining.
- New subcriterion to provide the opportunity for shifting time windows.
- Increased decline thresholds for Vulnerable.
- New threshold decline rates:

Sub-criteria	CR	EN	VU
A1,A3, A4	>30%	>50%	>80%
A2 (decline has ceased)	>50%	>70%	>90%

Figure A7.2 illustrates the principles behind the changes to Criterion A. The graph shows three kinds of decline. In (A) the population has declined rapidly but then stabilizes at a new much reduced level. This population would be assessed under the new Criterion A2 which has higher thresholds. Curves (B) and (C) show two different ways in which declines might proceed but where the decline is not halted. The thresholds in Criterion A1, A2 and A4 will apply to these.

Criterion B

The present area-based thresholds under Criterion B do not scale well across all organisms. Most of the time this is not a problem since criterion B is only intended to be applicable to species for which range area and distribution characteristics are the cause of threatened status, and not those for which population size and structure are measurable and relevant. However, the relatively large thresholds could lead to over-listing of some locally abundant, micro-endemic taxa. Scale of measurement under Area of Occupancy also has a strong influence on the resulting area.

Outcome

- A new structure for the criterion explicitly differentiates classifications made by Extent of Occurrence and Area of Occupancy.
- Additional guidelines on choosing scales for measurement of grid-based areas.

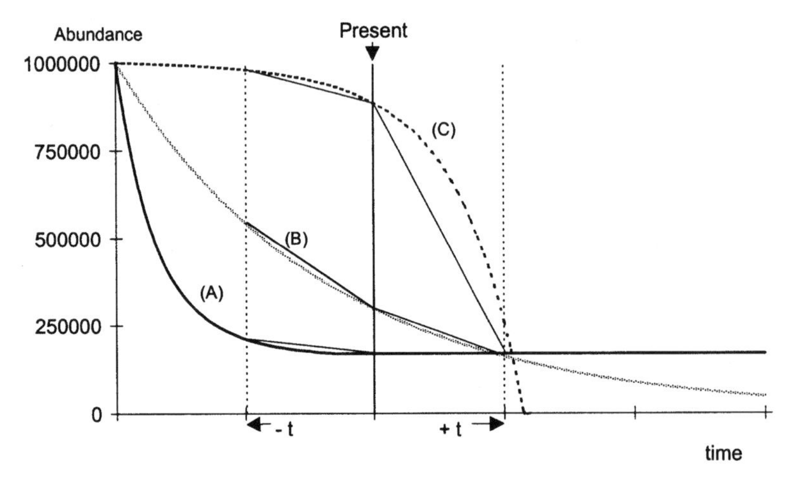

Figure A7.2. Patterns of population decline to which Criterion A might apply. In (A) the decline has ceased, in (B) the decline rate is reducing and in (C) the decline rate is increasing.

Criterion C

Under one of the qualifying sub-criteria, all individuals have to be in a single subpopulation. This is too exclusive and does not allow the listing of very skewed populations where a small number of mature individuals exist outside the main population.

Outcome

- A new form of C2(ii) to be more precautionary, allowing a small proportion of the population to be distinct.

Criterion D

Sub-criterion D2 under Vulnerable was intended to be used for species with very small distributions. However, the thresholds for area of occupancy and number of locations, although given as indicators, are frequently interpreted too literally. Some people have argued that the sub-criterion is too inclusive and results in massive over-listing, whereas others argue that it is too exclusive (for many marine species) and so is under-listing. The threats aspect needs to be emphasized more than the restricted distribution.

Outcome

- New wording of D2 under Vulnerable indicates that the quantitative thresholds are for guidance only here, to avoid over listing of micro-endemics.

Conclusion

There were some difficult and not fully agreed issues, which remained unresolved:

- Dealing with harvested species under Criterion A;
- Range areas measured using grids and the problems of scale of measurement; and
- Capping time scales in the past and future.

These issues will be addressed as much as possible in a comprehensive set of user guidelines. The revised IUCN Red List Categories and Criteria will come into force in 2001 and the aim is to keep this revised system stable for several iterations of the IUCN Red List. This stability will enable genuine changes in the status of species to be detected rather than to have such changes obscured by the constant modification of the criteria.

Georgina M. Mace
Institute of Zoology
Regent's Park
London
NW1 4RY
UK